COMPLETE
PRACTICAL MACHINIST

EMBRACING

LATHE WORK, VISE WORK, DRILLS AND DRILLING,
TAPS AND DIES, HARDENING AND TEMPERING,
THE MAKING AND USE OF TOOLS, TOOL
GRINDING, MARKING OUT WORK,

MACHINE TOOLS, ETC.

BY

JOSHUA ROSE, M. E.

395 ENGRAVINGS

19th EDITION

MERCHANT BOOKS

Copyright © 2006 Merchant Books

ISBN 1-933998-09-1

PREFACE TO THE NINETEENTH EDITION.

WHEN the first edition of this work was presented to the Mechanical Fraternity, the author made, in hopes and fears, his first bow to his fellow-craftsmen. Since then seventeen years have passed away and the encouragement he has received has led him to widen its scope in grateful recognition of the kind reception the work has received.

This new edition adds, the author believes, much value to the work, since it introduces to the reader the MACHINE TOOLS in which the cutting tools are used, whereas in all previous editions the cutting tools only were treated upon.

Care has been taken to keep the work up to date and to preserve all the familiar work-shop language that has led the practical workman to greet it so kindly.

JOSHUA ROSE.

DECEMBER 1, 1894.

CONTENTS.

CHAPTER I.

CUTTING TOOLS FOR LATHES AND PLANING MACHINES.

Importance of the Lathe; Steel of which Cutting Tools for Lathes, Planing Machines, etc., are made; Classification of Lathe-cutting Tools.............	25
Classification of Slide-rest Tools; The Forming of Cutting Tools; Illustration of the manner in which a Lathe Tool cuts the Metal; Principal consideration in determining the proper shape of a Cutting Tool..	26
Strain upon a Tool...................................	29
Rake in Tools	30
Principles determining the proper form of the cutting edge of a Tool....................................	33
Round-Nosed Tools....................................	36
Square-Nosed Tools...................................	37
Angles at which Tools become Cutters or Scrapers respectively.......................................	39
Effect of the diameter of the work and the rate of tool feed on the amount of clearance by the bottom rake or side rake of a Tool; Bearing of the height of the cutting edge of a Tool, with relation to the work, on its cutting qualities..........................	40
Practice of Sir Joseph Whitworth; Positions in which all Tools should be held.............................	41
Cutting off Parting or Grooving Tools.................	44

Side Tools for Iron.. 47
Front Tool for Brass Work............................. 50
Side Tool for Brass Work............................... 53
Special Forms of Lathe Tools.......................... 54
Tool Holders... 55
Woodbridge's patent Tools and Tool Holder........... 57

CHAPTER II.

Cutting Speed and Feed.

Meaning of the terms " Cutting Speed " and " Feed ",
    Planing Machines ; Great importance of " Feed "
    and " Speed " in Lathe Work............................. 64
"Feed " and "Speed " for various kinds of work 65
Tables of Cutting Speeds and Feeds ; Tables for Steel;
    for Wrought-iron... 69
For Cast-iron ; for Brass ; for Copper ; Speeds where the
    cuts are unusually long ones............................ 70

CHAPTER III.

Boring Tools for Lathe Work.

Standard Bits and Reamers ; The Shaping of Boring
    Tools for Lathe Work...................................... 71
Pressure on the cutting edge of a Tool, with Illustra-
    tion of the same... 72
Effect of the Application of the Top Rake or Lip to a
    Boring Tool.. 74
Shape for the corner of the cutting edge................ 75
Illustrations of the various forms of Boring Tools for
    ordinary use... 76
Boring Tool for heavy duty on Wrought-iron........... 77
Boring Tool for Brass.. 78
Easing a piece of bored Brass or Cast-iron Work, which
    fits too tight, with a half round Scraper............. 79
Boring Tools, with Illustrations........................... 80
Boring Tool Holders.. 82

CONTENTS.

CHAPTER IV.

SCREW-CUTTING TOOLS.

Cutting Surfaces of Lathe Tools for cutting Screws;
    Cutting the Pitch of a Screw which is very coarse..   84
The most accurate method of cutting small V Threads.   85
Tool for cutting an outside V Thread; Stout Tool for
    cutting coarse square Threads on Wrought-iron or
    Steel   86
Single pointed Tool for cutting an internal Thread;
    The three different shapes of V Threads in the
    United States—the sharp V Thread, the United
    States Standard, and the Whitworth Thread   87
Comparative ease in producing these several Threads..   88
Gauges for testing the angles of Threading Tools; Measuring the Diameter with the Calipers.............   90
Testing the Pitch of a Thread........................   91
Centre Gauge and Gauge for grinding and setting Screw
    Tools...   92
Experiments upon Targets representing ship's armor
    in which the bolts were found unable to resist the
    shock, and the remedy for the defect; To calculate
    the change Gear Wheels necessary to cut a given
    Pitch of Thread in a Lathe.......................   94
A simple or single-geared Screw-cutting Lathe.........   95
Compound or double-geared Screw-cutting Lathe......   97
Rule by which to find the number of Teeth in the
    Wheels to be placed on the Feed Screw............   98
Compound Gears common in small American Lathes...   100
Pitches of Threads used in France, and the method of
    finding the necessary change gears; To cut a Double
    Thread...   102
To cut a Treble Screw...............................   103
Hand Chasing.......................................   104
To make a Chaser...................................   107
Chaser used on Wrought-iron; An inside Chaser......   110
Views of an inside Chaser applied to a piece of work...   111
Uses of an inside Chaser; Cutting inside Threads......   113
General directions in cutting Threads.................   114

CHAPTER V.

LATHE DOGS, CARRIERS OR DRIVERS.

The Bent-tailed Dog, its objectionable feature, and how it may be obviated; The Clements Driver......... 116
Clamp Dog for rectangular or other work, not cylindrical at the driving end ; Form of Driver for driving Bolts; Adjustable Driver; Wood-turners' Spur centre..... 119
Screw Chuck for short wood work ; Mandrils or Arbors. 120
Centring Lathe Work ; Centre-grinding device......... 122
The quickest method of centring Lathe Work.......... 123
Centring Machine ; A centre-drilling attachment for Lathe Work...................................... 124
Combined Drill and Countersink for centre drilling; Combined Drill and Countersink, in which a small Twist-Drill is let into the Countersink ; Centre drilling by hand.. 125
Work requiring to be run very true ; A square centre ; To recentre work that has already been turned..... 126
Roughing out work which requires to be turned at both ends ; Finishing Lathe Work...................... 127
Emery Cloth and Paper...................... 128
Grinding Clamps for finishing work to gauge diameters ; Arbor for grinding out Bores............... 130
Lathe Chucks ; The three classes of Chucks ; Horton two-jawed Chuck ; Box body Chuck ; Three and four-jawed Chucks 131
The Sweetland Chuck................................. 133
Drill Chuck of the Russell Tool Co..................... 134
Chuck Dogs... 135

CHAPTER VI.

TURNING ECCENTRICS.

Chucking an Eccentric which has a hub or boss on one side only of its bore.. 136
Chucking an Eccentric having a large amount of throw upon it.. 138
Turning Crank... 140

CONTENTS.

To Chuck a Crosshead.................................. 141
Counterbalancing work................................. 143
Boring Links or Levers................................. 144
Turning Pistons and Rods.............................. 145
Piston Rings.. 146
Expanding Chuck for holding Piston Rings or similar work... 150

CHAPTER VII.

HAND-TURNING.

One of the most delicate and instructive branches of the Machinist's art 151
Far more instructive to a beginner than any other branch ; Chucking................................... 153
Roughing out ; The Graver............................. 154
Holding the Graver.................................... 156
The Heel Tool... 157
Hand-turning—Brass work.............................. 159
Scrapers.. 160

CHAPTER VIII.

DRILLING IN THE LATHE.

Work in which the Lathe is resorted to for drilling purposes... 164
Half-round Bits....................................... 166
Bit in which a Segment has been cut out to admit a Cutter... 167
Cutter and Bar designed for piercing holes out of the solid and of great depth ; Flat Drill to enlarge and true them out....................................... 168
Drill-holder.. 170
Reamers.. 171
Method of Grinding a Reamer.......................... 172
Importance of maintenance of the Reamer to Standard Diameter.. 174
Reamer which may be adjusted to size by moving its teeth ; Adjustable Reamer for very small work..... 175
Shell Reamers.. 176

CHAPTER IX.

BORING BARS.

Importance of the Boring Bar; Smaller sizes of Boring Bar usually simple parallel Mandrils.............. 178
Boring Bar Cutters requiring to be Adjustable......... 179
No Machine using a Boring Bar should be allowed to stop while the finishing cut is being taken.......... 180
A rude form of Head................................. 181
Position which Cutters should occupy towards the head or Body of the Bar................................ 183
Small Boring Bars.................................... 189

CHAPTER X.

SLOTTING MACHINE TOOLS.

Two Classes of Tools used in Slotting Machines........ 191
Tool for cutting a Half-round Groove, holding Bar and Short Tool....................................... 192
Knife Tool for heavy work............................ 193

CHAPTER XI.

TWIST DRILLS.

The Cutting Edges of Drills, with various examples.... 197
Testing Drills.. 199
The Flat Drill....................................... 201
To increase the Keenness of a Flat Drill; Feeding Drills... 202
The Farmer Lathe Drill; Experiments of Wm. Sellers & Co. with a Flat Drill......................... 204
Drilling Hard Metals................................. 205
Slotting or Keyway Drills............................ 206
Pin Drills... 211
Countersink Drills 212
Cutters.. 214

CONTENTS. 19

CHAPTER XII.

TOOL STEEL.

Cutting Tools for all machines should be made of
    hammered Steel 219
Forging Tools 220
Tool hardening and tempering 222
Hardening; To harden Springs 227
Case-hardening Wrought-iron 229
The wear of metal surfaces 230
Annealing or Softening 236
Mixtures of Metals 237

CHAPTER XIII.

TAPS AND DIES.

Forging of Taps 238
The Nut Tap 240
Taps having taper in the diameter of the bottom of the
    Thread; Proper taper for Hand and Machine
    Taps 241
Taps having Thread on the small end of the Taper;
    Turning the plain part of a Tap; Taps for use in
    holes to be tapped deeply; Finishing the Threads of
    a Tap 242
Flutes of small Taps; United States Standard for
    Threads, adopted by the Franklin Institute 243
English or Whitworth Standard; Hardening Taps 244
Taps of three and of four Flutes 246
The Whitworth, the Brown & Sharpe, and the Pratt &
    Whitney Taps; The position of the Square with re-
    lation to the cutting edges in Hand Taps 248
Adjustable Dies 250
Dies for use in Hand Stocks 251

CHAPTER XIV.

VISE-WORK—TOOLS.

Chisels; Flat Chisels 256

The Round-nosed Chisel; The Oil-groove Chisel...... 262
The Diamond-point Chisel............................. 263
The Side Chisel; Application of Chisels; Calipers...... 264
The Square... 267
The Scribing Block................................... 268
Files and Filing; Fitting Files to their handles; Selecting a File; Half-round Files...................... 269
Holding Files.. 270
Filing out Templates................................. 272
Scrapers and Scraping................................ 280
Vise Clamps.. 281
Vise-Work—Pening.................................... 282
Fitting Brasses to their boxes....................... 284
Fitting Link Motions................................. 285
Fitting Cylinders.................................... 288
Scraped Surfaces..................................... 297
To make a Surface Plate.............................. 301
To cut hard Saw Blades; To refit leaky Plugs to their Cocks.. 304
Refitting work by Shrinking it....................... 308
Steam and Water Joints............................... 313

CHAPTER XV.

FITTING CONNECTING RODS.

The mode of proceeding with the work................. 314
To get the length of a Connecting Rod; To ascertain when the Crank of a Horizontal Engine is upon its exact dead centre................................. 319
Fitting a Connecting Rod............................. 320
The Oil-Hole of a Connecting Rod; The Brasses or Side Rod.. 322
Drifts; Smooth and toothed or cutting Drifts......... 325
Reverse Keys... 329
Setting Line-shafting in Line........................ 331

CHAPTER XVI.

MILLING-MACHINES AND MILLING-TOOLS.

Importance of the Milling-machine.................... 338
Cutting out a Corrugated Surface..................... 339

Advantage of Milling-tools; To Mill the Side Faces of
    a Rod with Milling-bar and Cutters 340
Examples of work with the Milling-machine........... 341
The Side Faces of the Cutters........................ 343
Use of Milling-tools for cutting the Thread on Taps;
    Making the Milling-cutters....................... 344
Finishing the Cutter in the Lathe with an Emery
    Wheel 345
The Teeth of long Cutters............................ 346

CHAPTER XVII.

GRINDSTONE AND TOOL GRINDING.

Uses of Grindstones; Various kinds of Grindstones;
    Dry Grinding..................................... 347
Qualities of different Grindstones; Treatment of Grind-
    stones 348
To make a Grindstone run true; Accurate Grinding;
    Truing up a Grindstone for Tool Grinding......... 349
Objections to the intermittent truing of a Grindstone;
    Device for keeping a Grindstone continuously true 350
Face of a Grindstone for Flat Surfaces; Positions for
    holding Tools in Grinding........................ 352
A Feather Edge on a Tool............................. 353
A Device called a Rest 359

CHAPTER XVIII.

LINING OR MARKING OUT WORK.

Importance of Lining Out Work........................ 360
Principles involved in Marking Out Work; Qualities
    necessary for a Marker Out....................... 361
To mark an Ellipse................................... 363
To find Points through which the Curve of an Ellipse
    may be drawn 364
Tools employed by a Marker Out....................... 365
To divide a straight line into two equal parts; To di-
    vide a straight line into a number of equidistant
    points 367
Measuring Work to be Marked Out; Practice of Mark-
    ing Out.. 369

CONTENTS.

To Mark Off an Engine Guide Bar..................... 373
Use of the Compass Calipers in Marking Out Work... 376
Philosophy of Marking Out Holes in a certain manner. 377
Centrepunch Marks................................. 378
To Mark Off the Distance between the Centres of two Hubs of unequal height.......................... 379
Marking Holes at a right angle..................... 381
To Line Out a Double Eye.......................... 383
Marking Out an Eccentric........................... 389
Lining Out Connecting Rods......................... 396
To Mark Off Cylinder Ports and Steam Valves........ 406
Valve Seats.. 407
To Mark Out a Cone Pulley......................... 408

CHAPTER XIX.

MACHINE TOOLS.

A Machine Tool a Machine that Operates a Cutting Tool for Dressing the Work; The Lathe chief of all machine tools; The Shaping Machine being superseded by the Milling Machine; The simplest form of the Lathe, a Foot Lathe; The Foot Lathe illustrated and described 411
A Slide-Rest illustrated and described; Some objectionable features of Slide-Rests of this class........... 413
A Slide-Rest which obviates one of these objectionable features; A Foot-power Lathe by W. F. & John Barnes Co., Rockford, Ill.......................... 414
The Hendey-Norton Lathe, illustrated and described..... 416
The Tool Post; A Screw-cutting Engine Lathe constructed by the New Haven Manufacturing Co., of New Haven, Conn., illustrated and described............ 421
Warner & Swasey's Twenty-four Inch Universal Monitor Lathe illustrated and described; Friction Clutch of this Lathe.. 424
A Lathe in which the Spindle for the Dead-Centre is square and the Tail-Stock capable of being moved or fed crosswise of the Lathe; Turret Lathes; Turret Lathe manufactured by the Gisholt Machine Co., Madison, Wis., illustrated and described........... 426

CONTENTS.

The Screw Machine; A Special Lathe, in which the work is cut direct from the Bar without forging; The capacity of the Screw Machine greater than that of a Lathe.. 428
A Small Screw Machine illustrated and described 429
The Brown & Sharpe Manufacturing Co.'s Screw Machine illustrated and described; The Cutting-Off Machine, a form of Lathe used to cut Rods or Bars into exact lengths................................. 432
A Cutting-Off Machine constructed by the Hurlburt-Rogers Machine Co. illustrated and described; Shapers and Planers Machines, in which the tools cut in a straight line; A Shaping Machine illustrated and described................................... 435
The Tool-Carrying Head of this Shaping Machine illustrated and described 436
Fox's Pillar Shaping Machine illustrated and described; Tool-Carrying Ram or Bar Traversed in a Slide or Saddle, applied in Traverse Shapers, illustrated and described 440
Pillar Shaper of the Hendey Machine Co. illustrated and described 442
Morton's Shaping Machine illustrated and described 444
The Iron Planing Machine, or Iron Planer, illustrated and described 445
A Planer of a larger size illustrated and described 449
The Detrick & Harvey Machine Co.'s Open Side Planer illustrated and described 451
Milling Machines; Brown & Sharpe No. 0 Plain Milling Machine illustrated and described 453
Brown & Sharpe Universal Milling Machine, No. 3, illustrated and described................................ 455
A very solid and efficient Milling Machine of the Brown & Sharpe Manufacturing Co. illustrated and described. 457
Drilling Machines 458
Quint's Six-Spindle Turret Drilling Machine, for drilling and tapping small holes, illustrated and described; Slate's Sensitive Drilling Machine illustrated and described 459

CONTENTS.

Drilling Machine of Prentice Bros., Worcester, Mass., illustrated and described; The Feed Motions and how obtained.. 462
A Radial Drill, or Radial Drilling Machine, illustrated and described; Keyway Cutting Machines or Keyway Cutters 465
The Morton Manufacturing Co.'s Keyway Cutter illustrated and described... 466
The Slotting Machine.. 468
Slotting Machine of Wm. Sellers & Co., Philadelphia, illustrated and described 469
Bolt Cutting Machines or Bolt Cutters........................ 470
A National Bolt Cutter, in which the Jaws for holding the Bolt Heads are operated by the left-hand Hand Wheel, and the forward motion is operated by the Hand Wheels on the right, illustrated and described. 471

CHAPTER XX.

To Calculate the Speed of Wheels, Pulleys, etc.

CHAPTER XXI.

How to Set a Slide Valve.

Considerations in Setting a Slide Valve................. 477
Practical Operations in Setting a Valve................. 478

CHAPTER XXII.

Pumps.

Suction Pumps... 486
Force Pumps 488
Piston Pumps.. 489
A Plunger Pump.. 490
Efficiency of a Pump, how increased 491
Causes of loss of efficiency in Pumps.................... 493
Index 495

THE COMPLETE PRACTICAL MACHINIST.

CHAPTER I.

CUTTING TOOLS FOR LATHES AND PLANING MACHINES.

The lathe is the most important of all metal-cutting machines, or machine tools as they are termed, not only because of the comparative rapidity of its action, but also from the wide range and variety of operations that may be performed in it. He who is an expert lathe hand, or turner, will find but little difficulty in operating any other metal-cutting machine tool, because the methods of holding work and the shapes of the tools for other metal-cutting machines are similar, and are governed by the same principles as in the case of lathe work; hence, in this book all tools that are used in the lathe will be discussed under the head of lathe tools, notwithstanding that they may be also used in other machines.

Cutting tools for lathes, planing machines, etc., etc., are made of a special grade of cast-steel known as tool steel The tool is first forged to shape, and then hardened by heating it to a red heat and dipping it in water.

Lathe-cutting tools may be divided into two principal classes, viz., slide-rest tools and hand tools. The latter, however, have lost their former importance, because even small lathes are now provided with means to traverse the tools to the cut.

Slide-rest tools may be subdivided into two classes, those for inside or internal work, and those for external or outside work. They are designated from either the nature of the duty they perform, or from some characteristic peculiar to the tool itself. Thus a side tool is one that cuts upon a side or end face; a front tool is one that cuts in front; a spring tool is one that admits of deflection or spring, and so on.

In forming cutting tools it will be found that a very slight variation of shape, or of presentment to the work, causes appreciable difference in its cutting capacity, whether for smoothness or in taking off a quantity of metal. Furthermore, the shape of the tool must not only be varied for different kinds of metal, but also for extreme differences of hardness in the same kind of metal, more notably in the case of steel, some of which is almost as soft as wrought-iron, while the finer grades are exceedingly hard, especially when cut from the bar and not annealed or softened by being brought to a red heat and left to cool slowly. Cast-iron also is sometimes exceptionally hard, requiring a special shape of tool, while wrought-iron and brass vary but very little in their degree of hardness.

The manner in which a lathe tool cuts the metal from the work when fed along it is shown in Fig. 1, which represents a tool feeding a cut along a piece of wrought-iron, and it will be seen that the cutting comes off in a spiral. The diameter and the openness of this spiral depend entirely upon the shape of the tool, so that from the appearance of the cutting the quality of the tool may be judged.

The principles which govern the shape of tool necessary to cut a piece of metal under any given condition are general in their application; so that when these conditions are clearly understood it becomes a comparatively easy matter to shape a tool suitable for them.

The principal consideration in determining the proper

CUTTING TOOLS.

shape of a cutting tool, for use in a lathe or planer, is where it shall have the rake necessary to make it keen

Fig. 1.

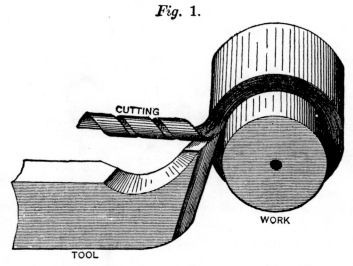

enough to cut well, and yet be kept as strong as possible; and this is governed, in a large degree, by the nature of the work on which it is to be used. It is always desirable, circumstances permitting, to place nearly all the rake or

Fig. 2.

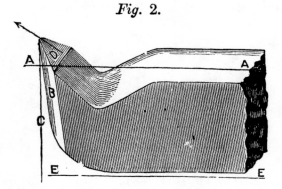

keenness on the top face of the tool, as shown in Fig. 2, in which D is the top face, and B the bottom one; lines A A and E E representing the level of the top and bottom of the tool steel, and C a line at a right angle to E, or what is the same thing, to A. The tool in Fig. 3 cor-

Fig. 3.

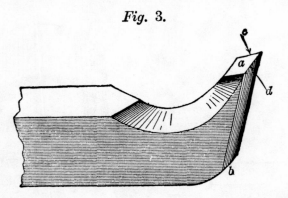

responds to that in Fig. 2 so far as its cutting qualifications are concerned, there being merely a slight difference in the forged shape, but not in the cutting edges. That shown in Fig. 3 is called a "diamond point," from the diamond shape of its top face, while that in Fig. 2 is called a "front tool;" the former being more suitable for small, and the latter for large work.

Referring now to the top face a, its angle or rake is its incline in the direction of the arrow in Fig. 4. In those

Fig. 4.

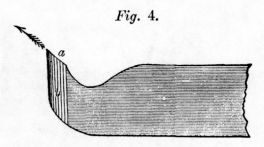

cases (to be hereafter specified) in which top rake is, from the nature of the work to be cut, impracticable, it must be taken off and the tool given the necessary keenness by increasing the rake or angle of the bottom or side faces in the direction shown in Fig. 5, in which letter *b* represents a side or bottom face of the tool, its amount of rake being denoted by its angle in the direction of the arrow.

Fig. 5.

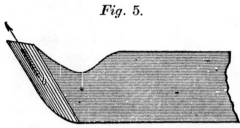

These top and side faces, taken one in conjunction with the other, form a wedge, and all machine tools are nothing more than cutting wedges, the duty performed by the respective faces depending, first, upon the keenness of the general outline of the top and bottom faces, and secondly, upon the position, relative to the work, in which the tool is held and applied.

The strain sustained by the top face is not alone that due to the severing of the metal, but that, in addition, which is exerted to break or curl the shaving, which would, if not obstructed by the top face, come off in a straight line, like a piece of cord being unrolled from a cylinder; but on coming into contact with the face of the tool (immediately after it has left the cutting edge), it is forced, by that face, out of the straight line and takes circular form of more or less diameter according to the amount of top rake possessed by the tool. The direction of the whole strain upon the top face is at a right angle to it, as denoted in Fig. 6 by the line D, *d* being the work, B the tool, and C the shaving. It will be readily perceived, then, that if a tool possessing so much top rake is held far out from the tool post or clamp, or is slight in body, any springing of the body of the tool, arising from the pressure due to the cut, will cause the tool point to take a deeper cut, and that the tendency of the strain upon the top face is to draw the tool deeper into its cut. A plain cut (either inside or outside) admits of the use of a maximum of top rake and of a minimum of bottom rake in all cases when the tool is not liable to spring.

Were the strain upon the tool equal in force at all times during the cut, the spring would also be equal, and the cut, therefore, a smooth one; but in taking a first cut, there may be, and usually is, more metal to be cut off the work in one place than in another; besides which there are inequalities in the texture of the metal, so that when the harder parts come into contact with the tool, it springs more and cuts deeper than it does when cutting the softer parts, and therefore leaves the face of the work uneven.

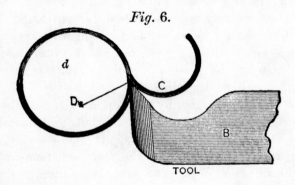

Fig. 6.

The main duty performed by the bottom face of the tool is to support the cutting edge, and the amount of rake it possesses is not, under ordinary circumstances, of very great consequence, so that it be sufficient to well clear, and not rub against the work. It is always desirable, however, to give it as little rake, over and above clearance, as possible, to avoid weakening the cutting part of the tool.

When, in consequence of the top face having but very little rake, it becomes necessary to make the general outline of the tool keen by the application of the maximum of side or bottom rake, the tool becomes proportionately weak, as is shown in Fig. 7; in which *a* represents the work, B the tool, *c* the shaving, and D the direction of the strain placed upon the top face of the shaving, from which, it will be noted, that the cutting edge is comparatively weak, and hence, liable to break.

Fig. 7.

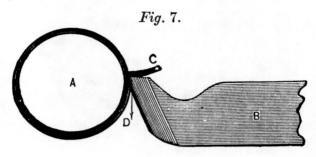

It follows, then, that if two tools are placed in position to take an equal cut off similar work, that which possesses the most top rake, while receiving the least strain from the shaving, receives it in a direction the most likely to spring it into its cut. It must not, therefore, be used upon any work having a tendency to draw the tool in, nor upon work to perform which the tool must stand far out from the tool post, for in either case it will spring into its cut.

Especially is this likely to occur if the cut has a break in it with a sharply defined edge, such, for example, as turning a shaft with a dovetailed groove in it. Taking all these considerations into account, we arrive at the tool shown in Fig. 8, as representing the most desirable amount of top and bottom rake for ordinary purposes on light work; such a tool is not, however, adapted to taking very heavy cuts, for which duty the top face of the tool is given what is termed side rake.

Fig. 8.

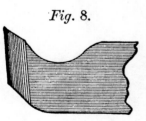

Fig. 9 represents a tool having a maximum of side rake, and therefore designed for very heavy duty, and to be held as close to the tool post as possible. The amount of power required to feed a lathe or other tool into its cut, at the

Fig. 9.

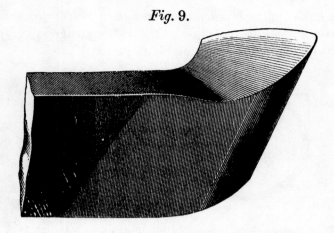

same time that the tool is cutting, is considerable when a heavy cut is being taken; and the object of side rake is not only to make the tool more keen without sacrificing its strength, but to relieve the feed screw or gearing of part of this strain by giving the tool a tendency to feed along and into its cut, which is accomplished by side rake, thus:

Fig. 10.

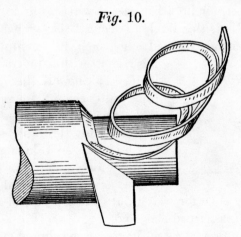

Suppose Fig. 10 to represent a tool having side rake its feed being to the left and the pressure of the shaving will be more sideways. It has in fact followed the direction of the rake, decreasing its tendency to run, or spring, in (as shown in Fig. 6), with a corresponding gain in the above-mentioned inclination to feed itself along, or into, its lateral cut.

When side rake is called into use, a corresponding amount of front rake must be dispensed with, or its tendency to feed itself becomes so great that it will swing round, using the tool post as a centre, and (feeding rapidly into the cut) spring in and break from the undue pressure, particularly if the lathe or machine has any play in the slides. So much side rake may be given to a tool that it will feed itself without the aid of any feed motion, for the force required to bend the shaving (in heavy cuts only) will react upon the tool, forcing it up and into its cut, while the amount of bottom rake, or clearance as it is sometimes called, may be made just sufficient to permit the tool to enter its cut to the required thickness of shaving or feed and no more; and it will, after the cut is once begun, feed itself and stop of itself when the cut is over. But to grind a tool to this exactitude is too delicate an operation for ordinary practice. The experiment has, however, been successfully tried; but it was found necessary to have the slides of the lathe very nicely adjusted, and to take up the lost motion in the cross-feed screw.

For roughing out and for long continuous cuts, this tool is the best that can be used; because it presents a keen cutting edge to the metal, and the cutting edge receives the maximum of support from the steel beneath or behind it. It receives less strain from the shaving than any other; and will, in consequence of these virtues combined, take a heavier cut, and stand it longer, than any other tool; but it is not so good for taking a finishing cut as one having front rake, as shown in Fig. 8.

Having determined the position of the requisite rake, the next consideration is that of the proper form of the cutting edge, the main principles of which are as follows:

Fig. 11 is a side and Fig. 12 an end view of a tool having a combination of front and side rake, and it will be understood from what has been already said that the front rake will cause the pressure of the shaving or cutting

Fig. 11.

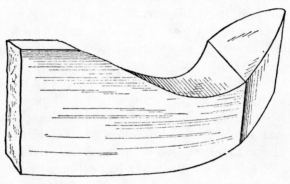

to pull the tool forward and into its cut, while the side rake will act to pull the tool along in the direction of its feed traverse. Now, when the tool is first moved to put on the cut, the cross-feed screw in moving the tool towards the lathe centre will bear on the sides of the threads nearest to the back of the lathe, and if there is any play or lost motion between the threads of the cross-feed screw and its nut, then so soon as the tool edge meets the work the front rake will cause it (under a heavy cut) to move inwards to whatever amount the play in the screw and its nut will permit it to go. Similarly the carriage-moving mechanism will not move the carriage until all the lost motion in its parts is taken up, and when the tool meets the work the strain of a heavy cut will be sufficient from the side rake to pull the tool forward in the direction of the feed, and these two motions

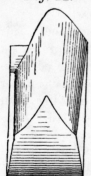

Fig. 12.

CUTTING TOOLS. 35

combined will cause the tool to dig in and probably break. To avoid this difficulty we may adopt two methods:

Fig. 13.

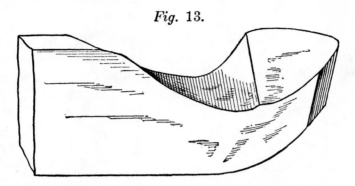

first we may take off the front rake, leaving the side rake intact as in the side view, Fig. 13, and the end view, Fig. 14, which will prevent the tendency of the tool to move in towards the lathe centre, and next we may set the tool in too far and be winding it outwards with the cross-feed screw at the time the tool-edge first strikes the cut. A better plan, however, is to give the top face negative top rake, as in Fig. 15, from A to B, in which case the pressure of the cut or rather of the cutting on the top face of the tool will act to a great extent to force the tool back and away from

Fig. 14.

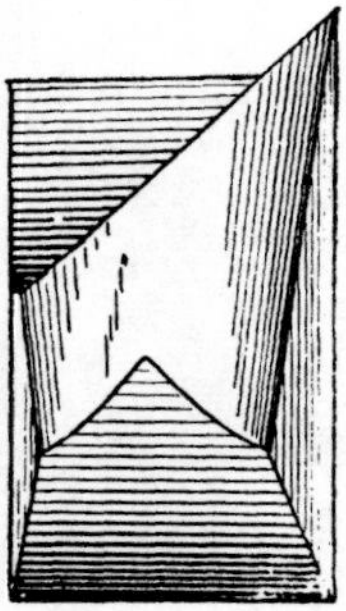

Fig. 15.

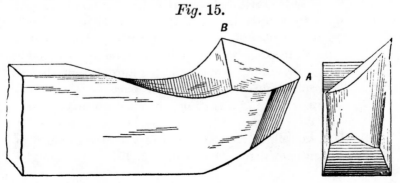

the work, and it will therefore take its cut gradually and easily.

ROUND NOSED TOOLS.

Round nosed tools, such as shown in Fig. 16, have a greater length of cutting edge to them (the depth of the cuts being equal) than the more pointed ones, such as was shown in Fig. 3, and as a result they receive more strain from, and hence are more liable to run into or out from, the cut. If sufficient rake is given to the tool to obviate this defect, it will, under a heavy cut, spring in. It is, however, well adapted to cutting out curves, or taking finishing cuts on wrought-iron work, which is so strong and stiff as not to spring away from it, because it can be used with a coarse feed without leaving deep or rough tool or feed marks; it should, however, always be used with a slow speed. On coming into contact with the scale or skin of the metal, in case the work will not true up, it is liable to spring away from its cut and therefore to cut deeper into the softer than into the harder parts of the metal. The angles or sides of a cutting tool must not of necessity be quite flat (unless for use on slight work, as rods or spindles), but slightly curved, and in all cases rounded at the point, as in the tool shown in Fig. 17. If the angles

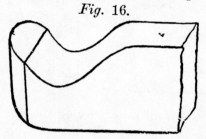

Fig. 16.

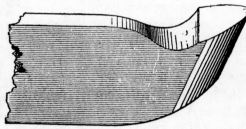

Fig. 17.

were left flat and the point sharp, the tool would leave deep and ragged feed marks; the extreme point, wearing away quickly, would soon render the tool too dull for use, and the point would be apt to break.

For finishing small wrought-iron work it should be ground, as shown in Fig. 18, being far preferable to the

Fig. 18.

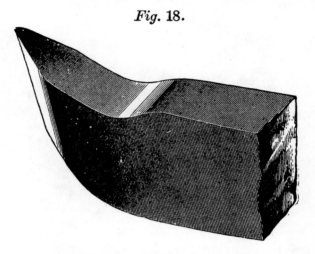

square-nosed finishing tools sometimes used for that purpose, since such tools do not turn true but follow the texture of the metal, cutting deepest in the softer parts, especially when the tool edge becomes the least dulled from use. It should be used with a quick speed and fine feed. On turning work of one inch and less in diameter, it is an excellent roughing tool, and with the addition of a little side rake is, for work of two inches and less diameter, as good a tool for roughing out as any that can be used.

SQUARE-NOSED TOOLS.

Square-nosed tools, such as shown in Fig. 19, should never be used upon wrought-iron, steel, or brass, for a broad cutting surface running parallel with the line of feed will always, upon either of these metals, cause the tool point to spring into the softer parts and to spring away from the harder parts, and, if the tool is liable to spring, in most cases, to dig into the work. Upon cast-iron work, however, such a tool will work to great advantage either for roughing out or finishing. It should be set so that its square nose is placed quite parallel with the work;

Fig. 19.

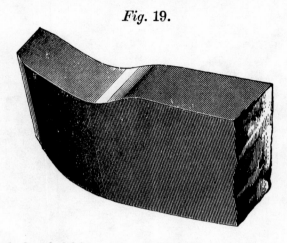

the feed for finishing purposes being almost as broad as the nose of the tool itself, or say three revolutions of the lathe per inch of tool travel. It should be fed very evenly, because all tools possessing a broad cutting surface are subservient to spring, which spring is, in this case, in a direction to deepen the cut; so that, if more cut is taken at one revolution or stroke than at another, the one cut will be deeper than the other. They are likewise liable to jar or tremble, the only remedy for which is to grind away some of the cutting face or edge, making it narrower. For taking finishing cuts on cast-iron, more top rake may be given to the tool than is employed to rough it out, unless the metal to be cut is very hard; else the metal will be found, upon inspection, to have numerous small holes on the face that has been cut, appearing as though it were very porous. This occurs because the tool has not cut keenly enough, and has broken the grain of the metal out a little in advance of the cut, in consequence of an undue pressure sustained by the metal at the moment of its being severed by the tool edge.

The angle of the top and of the bottom face of a tool does not determine whether it shall act as a scraping or cutting tool, but merely affects its capability of withstanding the strain and wear due to severing the metal which it cuts. Nor is there any definite angle at which the top face, B, to the work converts the edge from a cutting to a scraping one. A general idea may, however, be obtained by reference to Fig. 20, the line A being in each case one drawn from the centre of the work to the point of contact between the tool edge and the work, C being the work, and B the tool. It will be observed that the angle of the top face of the tool varies in each case with the line A. In

Fig. 20.

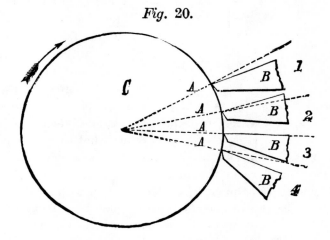

position 1, the tool is a cutting one; in 2, it is a scraper; in 3, it is a tool which is a cutter and scraper combined, since it will actually perform both functions at one and the same time; and in 4, it is a good cutting tool, the shapes and angles of the tools being the same in each case.

It may now be shown that the amount of clearance given by the bottom rake or side rake of a tool depends upon the diameter of the work and the rate of tool feed. In Fig. 21, for example, we have three tools in positions marked respectively 1, 2 and 3, the amount of rake being such as will give 5 degrees of clearance from the cut in each case. The lines A are at a right angle to the work axis, and we perceive that in position 1 the tool has $8\frac{1}{2}$ degrees of angle from A, while in position 2 it has $10\frac{1}{2}$, and in position 3, 15 degrees of angle from A. This occurs because the angle of the cut to the work axis is greater in proportion as the diameter of the work is less. It is obvious, therefore, that to give equal clearance, the bottom rake must vary for every different diameter of work or rate of feed.

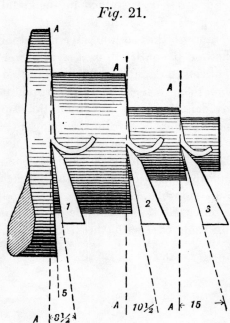

Fig. 21.

The height of the cutting edge of a tool with relation to the work has an important bearing upon its cutting qualifications, since it affects the angle of the top face, as may be seen from Fig. 22, in which the tool being placed extremely high, cutting is bent but little out of the straight line. It is obvious, however, that if the conditions are such as to cause the tool to bend under the pressure of the cut and more at one part of a work revolution than at another, then the work would be turned out of true, or out of round, as it is sometimes termed.

LATHE AND MACHINE TOOLS. 41

In Figs. 23 and 24 are two tools of the same shape, but placed at different heights, and it is seen from the dotted

Fig. 22.

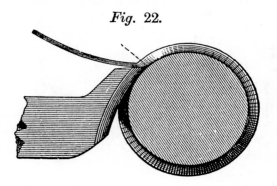

lines that the lower the tool the longer the line of resistance of the metal to the cutting action of the tool.

Sir Joseph Whitworth designs his lathes so that the tool requires to be forged as in Fig. 25 to bring its cutting edge level with the axis of the work W, so that if the tool bends under the cut pressure (which it will do to some extent, however rigidly it may be held), it will move in a direction having a minimum of effect upon the roundness of the work. Thus let R represent the lathe rest and A the fulcrum or point off which the tool springs or bends, and the arrow will represent the direction of tool spring.

Fig. 23.

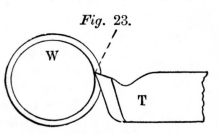

Fig. 24.

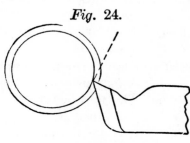

It follows therefore that all tools should be held so that their cutting edges are as near the tool post as possible, so as to avoid their spring-

4*

ing, and to check as far as possible their giving way to the cut, in consequence of the play there may be in the slides of the tool rest; but if, from the nature of the work

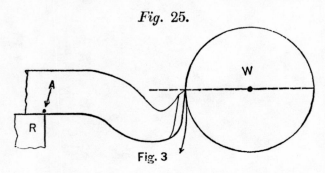

Fig. 25.

Fig. 3

to be performed, the tool must of necessity stand out far from the tool post, we should give the tool but little top rake, and be sure not to place it above the horizontal centre of the work.

The cutting tools for planing machines are subservient to the same spring, but the effect is less upon the work, because if, in consequence of the spring or deflection, a lathe tool approaches the work axis say one thousandth of an inch, then the work will be turned smaller in diameter to twice this amount, whereas in a planer tool the work will only be affected to the same amount as the tool deflects. Fig. 26 represents a planer tool, the point A being the fulcrum off which it springs, and the arrow the direction in which the spring occurs. This may be remedied so far as its effect upon the work is concerned by shaping the tool so that its cutting edge falls vertically under the centre of the tool steel, as denoted by the dotted line in Fig. 27, in which case the tool will cut very smoothly. In

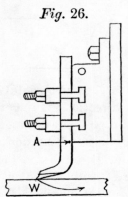

Fig. 26.

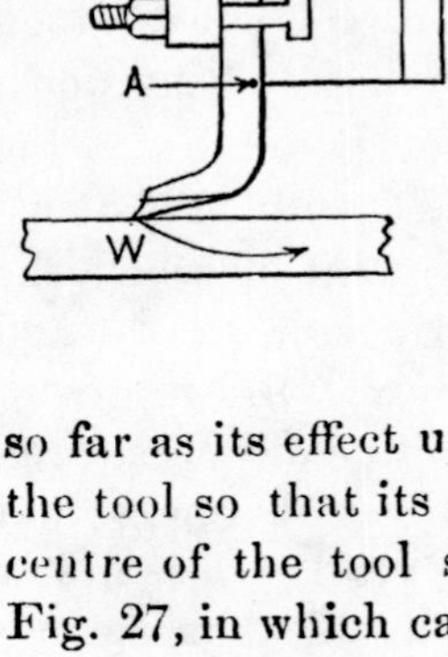

LATHE AND MACHINE TOOLS.

other practice, and especially for broad finishing tools for cast-iron, the cutting edge is made to stand about

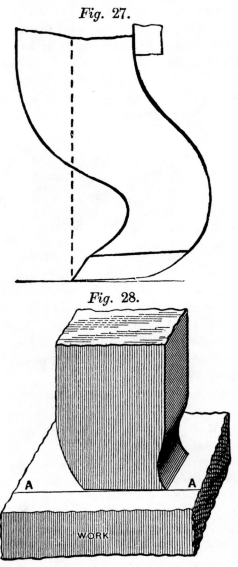

Fig. 27.

Fig. 28.

level with the top of the tool steel, the bottom clearance being a minimum as shown by the line A in Fig. 28.

CUTTING OFF PARTING OR GROOVING TOOLS.

Tools that are necessarily slight in form, especially those for use in a planer, are more subservient to the evil effects of spring than those of stouter body; and in light planers, when the tool springs in, the table will sometimes lift up, and the machine become locked, the cut being too deep for the belt to drive. The tool most subservient to spring is the parting or grooving tool shown in Fig. 29, which,

having a square nose and a broad cutting surface placed parallel to the travel of the cut, and requiring at times to be slight in body, combines all the elements which predispose a tool to spring, to obviate which, it should be placed at or a little below the centre, if used in a lathe under disadvantageous conditions, and bent similarly to the tool shown in Fig. 27, if for use in a planer, unless under favorable conditions.

The point is made thicker to give clearance to the sides, so that it will only cut at the end, and the breadth is left wider than other parts to compensate in some measure for the lack of substance in the thickness.

For use on wrought-iron or steel, when the tool is very thin, or when it requires to enter the metal to an unusual depth, or requires to stand far out from the tool post, the tool should be made as shown in Fig. 32, which will obviate the necessity of bending the body of the tool, and prevent it from the digging in and breaking off so common under those conditions. When used upon wrought-iron or steel, the cutting point should be **freely** supplied with oil or soapy water.

This tool is obviously fed endways into its cut, as shown in Fig. 30, and for grooving purposes **cuts better if it**

Fig. 30.

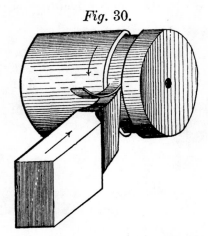

has its cutting edge set above the work axis, but if the work is to be entirely severed, the cutting edge must be set level with the work axis, and the feed be very fine towards the last. For brass work the top face should be ground depressed towards the point as shown in Fig. 31, which

Fig. 31.

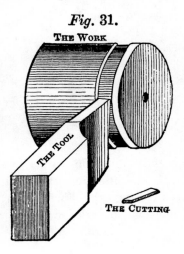

will cause it to cut more smoothly and avoid the jarring or chattering which is otherwise very apt to occur.

The spring tool, shown in Fig. 33, is especially adapted to finishing sweeps curves, and round or hollow corners, and may be used with equal advantage on any kind of

Fig. 32.

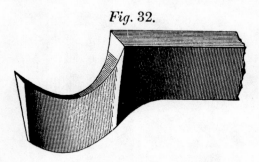

metal whatever, performing its duty more perfectly than any other form of tool could, since the conditions under which it operates, that is, a very broad cutting surface, would cause any other tool to dig into the work. The spring tool, however will spring rather away from than

Fig. 33.

into its cut, the only objection to its use being that in consequence of this qualification it is apt to spring into the softer and away from the harder parts of the metal. Its capability, however, to take a broad surface of cut, when the cutting edge stands a great way out from the tool post, renders its use for some work imperative as a finishing tool, while under ordinary conditions it will perform its duty sufficiently accurately for all practical purposes. As illustrated, its top face has a little rake to fit it for use on wrought-iron; for use on brass and cast-iron the top face should have negative top rake. In cases where the conditions render it liable to spring, the horizontal level of the top face may be made even with the bottom face of the body of the tool, or the body of the tool may be bent for the reasons explained by Figs. 25, 26, and 27. and the accompanying explanations.

The top face of a spring tool should be filed up very smoothly before being hardened, and it should never be ground upon that face. The bevel in the direction of the arrow should be less for cast-iron and brass than for other metals, but should in no case be excessive, whatever the inclination of the top face may be. The bend of the tool should be left soft, the cutting face being hardened to a straw-color for stout tools, and a brown for slight ones. The face denoted by the arrow should, after grinding, be smoothed with an oilstone. For use on steel and wrought-iron, it should be freely supplied with either soapy or other water; and for finishing cast-iron, such water may also be used; and that metal will cut as clean and as polished as wrought-iron, providing the speed at which it is cut is a *very slow one.* When this tool is to be used a very long way out from the tool post, the wooden wedge, shown driven in the bend, should be taken out.

SIDE TOOLS FOR IRON.

Side tools for iron are subject to all the principles already explained as governing the shapes of front tools, and differ from them only in the fact that the cutting end of the tool is bent around to enable the cutting edge on one side to cut a face on the work which stands at right angles with the straight cut. A front tool is used to take the straight cut nearly up to the shoulder; then a side tool is introduced to take out the corner and cut the side face.

Fig. 34.

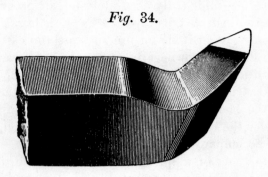

A side tool, whose cutting end is bent to the left, as in Fig. 34, is called a left-handed side tool; and one which is bent to the right, a right-handed side tool. The cutting edges should form an acute angle, so that, when the point of the tool is cutting out a corner, either the point only or one edge is cutting at a time; for if both of the edges cut at once, the strain upon the tool causes it to spring in.

For heavy work it may be made more round-nosed, and allowed to cut all round the curve, and with a coarse feed. It is also an excellent tool for roughing out sweeps or curves; and for use on small short bolts, it may be used on the parallel part as well as under the head.

For taking out a corner or fillet in slight work, which is liable to spring from the pressure of the cut—the point must be rounded very little, and the fillet be shaped by operating the straight and cross feed of the lathe. It is made right or left-handed by bending it in the required direction, that shown being a left-handed one.

The form of side tool shown in Fig. 34 is that most desirable for all small work where it can be got in; and in the event of a side face being very hard, it possesses the advantage that the point of the tool may be made to enter the cut first, and, cutting beneath the hard skin, fracture it off without cutting it, the pressure of the shaving on the tool keeping the latter to its cut, as shown in Fig. 35.

Fig. 35.

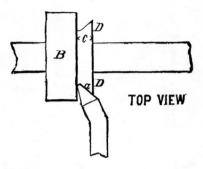

TOP VIEW

a is the cutting part of the tool; B is a shaft with a collar on it; *c* is the side cut being taken off the collar, and D is the face, supposed to be hard. The cut is here shown as being commenced from the largest diameter of the collar, and being fed inwards so that the point of the tool may cut well beneath the hard face D, and so that the pressure of the cut on the tool may keep it to its cut, as already explained; but the tool will cut equally as advantageously if the cut is commenced at the smallest diameter of the collar and fed outwards, if the skin, D, is not unusually hard.

Fig. 36.

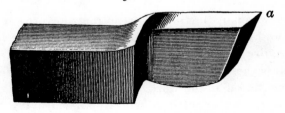

For cutting down side faces where there is but little room for the tool to pass, the tool shown in Fig. 36 is used, *a* being the cutting edge. Not much clearance is required on the side face of this tool, its keenness being given by the angle of the top face. Fig. 37 represents an end view of the tool at work. When the work is of small diameter, the cutting edge may be ground straight and set at a right angle to the work axis, so that the tool may be set in its full depth and fed laterally. In the case of work of large diameter, however, the tool should be formed and set as in Fig. 38, the cut being taken at the point E and fed from the circumference to the centre of the work. In some cases this tool is forged as in Fig. 38, being thicker at the bottom; this, however, is only advantageous when heavy cuts are taken and greater strength is therefore necessary. For small work, such as facing under the heads of bolts, the cutting end is bent at an angle, so that the tool will clear the work driver when set as in Fig. 39.

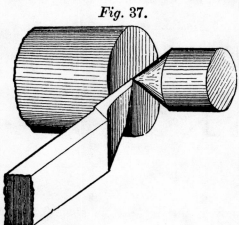

Fig. 37.

FRONT TOOL FOR BRASS WORK.

The main distinction between tools for use on iron or steel, and those for use on brass work is, that the latter do not require any top rake. Fig. 40 represents a front tool for brass, and Fig. 41 shows the manner in which the cuttings fly off if the work is run as fast as it should be. The distance the cuttings will fly after leaving the tool gives a very good indication of the efficiency of the tool; ordinary composition brass flying 15 or 20 feet. The front tool is a complete master tool, filling every qualification for all plain outside work, both for roughing out and finishing. For very light work, or when the tool must be held far out from the tool-post, it may be given

Fig. 38.

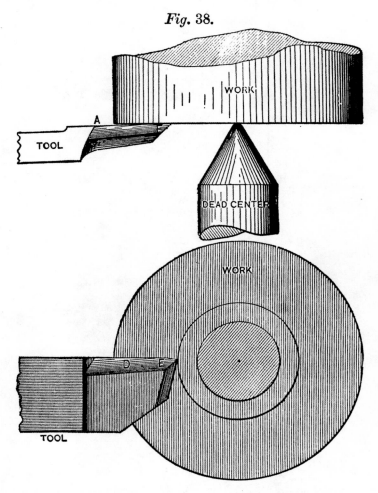

a little more rake on the bottom or side faces; while for finishing, the point may be more rounded and used with a coarser feed, providing the tool is rigid and not liable to spring. When held far out from the tool-post, the side

Fig. 39.

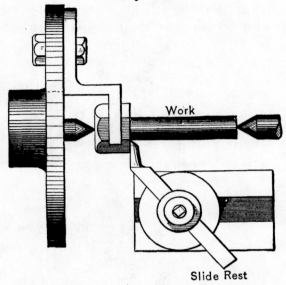

Fig. 40.

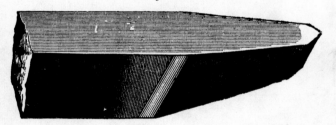

Fig. 41.

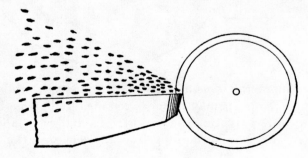

faces may be ground keener, and the top face have negative top rake—that is to say, some of the rake may be ground off the top face, and more given to the bottom or side faces; under such conditions, also, the cutting surface on the point of the tool may be reduced as small as convenient, so as to avoid the liability to spring. When ground round-nosed and smoothed with an oilstone, this tool gives a true and excellent finish to plain work.

SIDE TOOL FOR BRASS WORK.

The best side tool for brass is that shown in Fig. 42. It requires little or no top rake, and but little side or bottom rake unless used upon very slight work, or used

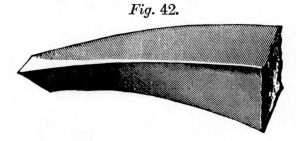

Fig. 42.

under conditions rendering it liable to spring. For taking out corners, and for turning out recesses which do not pass entirely through the metal, it has no equal. When it is held far out from the tool-post, it should have the top face bevelled off, at an angle of which the cutting part is the lowest, which will thus prevent it from jarring or chattering, and from springing into the work. In grinding it, only the end should be ground, so that the curve of the side of the tool—which is intended to allow the body of the tool to clear the shoulder or flange of the work—shall be preserved.

It will take a parallel cut, provided the corner is slightly rounded, as easily and well as a side cut; and for small work, can be used to advantage for both purposes.

It is a far better tool than those bent around at the end after the manner of a boring tool, being easier to forge, easier to grind, and not so liable to either spring, jar, or chatter.

If a tool for use on brass be made too keen, it will give the surface of the brass a mottled appearance, the color appearing lighter in patches. Furthermore, the face of the cut will appear jarred or chattered, and the cutting must be performed at a slower speed and feed.

SPECIAL FORMS OF LATHE TOOLS.

When it is required that a lathe tool shall produce upon a number of pieces of work, a sweep curve or fillet that must be of the same form in all the pieces, the difficulty arises that it is troublesome to grind the tool without altering its shape. This, however, may be avoided by the use of circular cutters, such as shown in Fig. 43, in

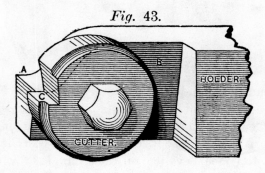

Fig. 43.

which the cutter is cylindrical, and has at the corner C the reverse of the curve it is required to produce on the work. The cutter is sharpened by grinding the horizontal face C which is set level with the face A of the holder, this face being made level with the line of lathe centres, when the holder lies horizontal in the tool clamp or tool post. Clearance is obtained partly by reason of the curve of the cutter, and partly by inclining the face B against which the cutter is bolted.

Figs. 44 and 45 show an application of a cutter of this kind for cutting a thread, it being obvious that to cut a right-hand thread on the work, a left-hand one must be provided on the cutter.

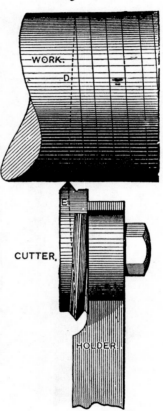

Fig. 44.

TOOL-HOLDERS.

To avoid the trouble of forging tools and to give greater rigidity to the tool when it requires to stand far out from the tool-post or clamp, various forms of tool-holders are employed. Thus in Fig. 46 is a tool-holder consisting of a bar A having a hub or boss H, an end view of the same being given in Figs. 47 and 48, in which it is seen that the tool is composed of a triangular piece of steel held between two pieces that fit inside the hub of the holder, and are clamped against the tool by a set screw. The tool and these pieces may be revolved in the hub to set the tool at any required angle as in screw cutting.

Fig. 49 represents a tool-holder H, having a clamp C secured by the bolt B, and having a feather at f, to hold the clamp horizontal when bolt B is loosened. The tool T has a groove on its side receiving a feather K, which is fast in the holder, and therefore holds the tool at a constant angle. At S is a screw threaded into the edge of the tool T, so that by operating this screw the height of the tool may be regulated. By this means the tool may

Fig. 45.

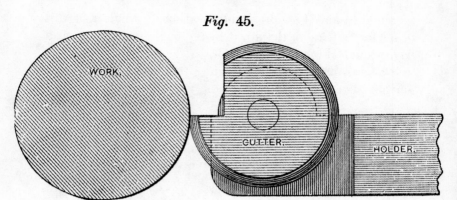

Fig. 46.

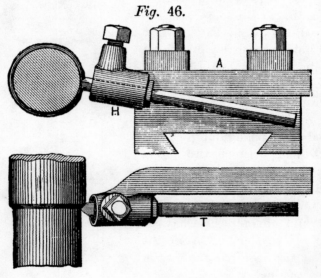

Fig. 47.

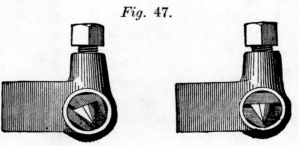

LATHE AND MACHINE TOOLS. 57

be made to any required shape, and as it is sharpened by grinding the top face only, this shape will be maintained as long as the tool lasts.

Figs. 50, 51, 52 and 53 represent Woodbridge's patent tools and toolholder. The tools consist of straight pieces of steel bevelled at the top to give a certain amount of side rake, the only grinding required to sharpen them being on the

Fig. 48.

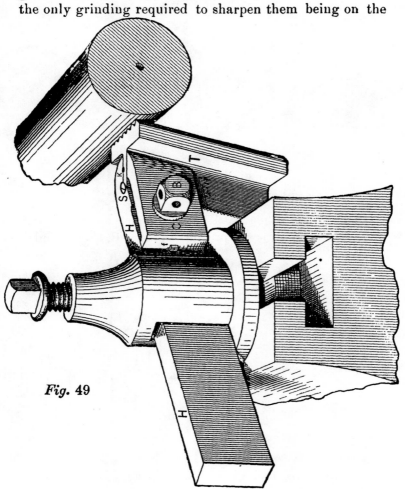

Fig. 49

end face. The tools are hardened throughout, and hence, require neither forging nor tempering. For left-hand tools the holder is turned end for end, so that the tool may be sustained by the holder as near to the cutting edge as possible, as is shown in Fig. 52, which represents a right and

Fig. 50.

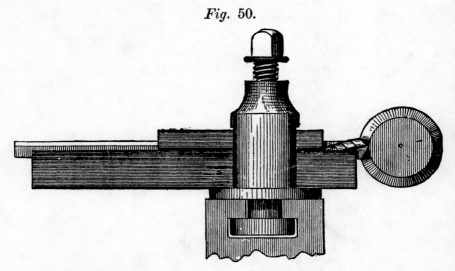

a left-hand tool in place. The cap which sets over the tool and receives the set screw pressure binds at B, Fig. 53, only,

Fig. 51.

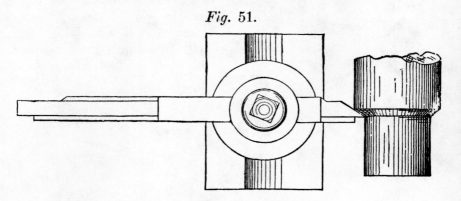

and the seat A in the holder is at an angle so as to give the side J of the tool the necessary clearance.

Fig. 54 represents a combined cutting off tool-holder and steadying device which is intended for cutting from rods pieces of an exact length. The holder is secured

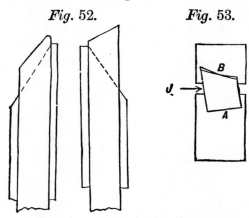

Fig. 52. Fig. 53.

in the tool-post, and has three screws which are set to steady the rod (which of course passes through them).

Fig. 54.

On the side of the holder is a slideway carrying a slide to which the cutting-off tool is fixed, being fed to its cut by the crank handle shown.

Fig. 55 represents a cutting-off device in which one leg or arm carries steadying pieces adjusted by means of the thumb

Fig. 55.

screw shown, while the other arm carries a gauge and a pivoted piece carrying the cutting tool, which is fed to the cut by the second arm, which is pivoted to the first one. This affords a very ready means of cutting off pieces for small work, since it squares the work ends at the same time that it cuts it off.

Fig. 56 represents a tool-holder for a shaping machine,

Fig. 56.

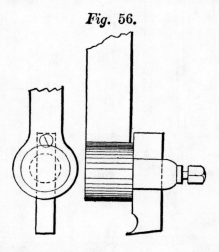

the tool being carried in a tool-post at the back of the holder so that it is pulled rather than pushed to its cut, and is not so liable to dip into the work from the spring or deflection. In place of a tool-post the tool or cutter may for curves, fillets, etc., be bolted direct to the holder, as in Fig. 57, B representing the cutter. Fig. 58 represents another form of tool-holder for shaping or planing machines, the tool being carried in a pivoted tool-post at the end of the holder, so that it may be swung to the right or left as may be required; a side and a front view of this tool and holder in place upon the planing machine sliding head, is shown in Fig. 59. The tool is set at an angle to give it front rake. The objection to this form is that the tool is partly hidden by the tool-post. Applications of this tool-holder are shown in Fig. 60, the direction of the feed being denoted by the arrows. Fig. 61 represents an

Fig. 57.

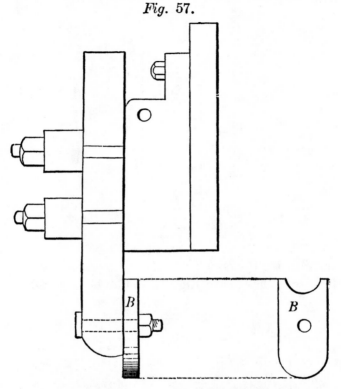

exceedingly useful form of tool-holder for planing machine tools, applications of its use being shown as follows.

Figs. 62, 63, 64 and 65 show the application of such a

Fig. 58.

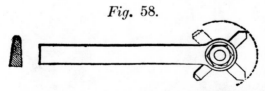

holder and tools to the cutting out of a T-shaped groove from solid metal. A grooving tool first cuts two grooves, as shown in Fig. 62, where one groove is shown finished and the tool is in operation on the second one. The next operation would be to cut out the metal between these two

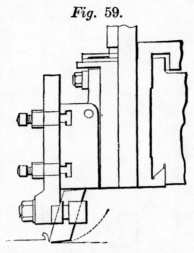

Fig. 59.

grooves, using the same tool. In the absence of the holder a bent tool, such as in Fig. 63, would be required to cut

Fig. 60.

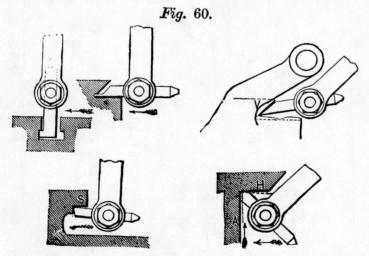

these grooves, and the tool being less rigid would not be able to carry so heavy a feed, nor would it produce so smooth a cut. This groove being finished, a tool having a single bend may be used to cut out the enlargement on one side, as shown in Fig. 64, carrying down a groove at each end, and then cutting out the metal left between them. In the absence of the holder the tool would require to have two bends, as shown in Fig. 65, and being made of large steel would be more difficult to forge and troublesome to use on account of its liability to spring and bend under the pressure of the cut, whereas, on account of the stiffness of the holder, its tools may be made of small pieces of steel.

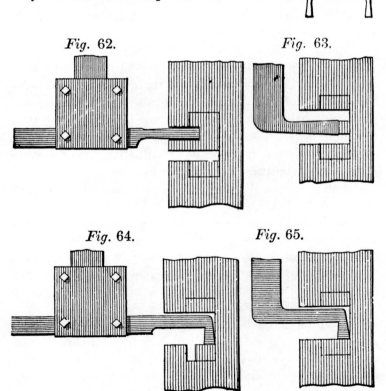

Fig. 61.

Fig. 62.   Fig. 63.

Fig. 64.   Fig. 65.

CHAPTER II.

CUTTING SPEED AND FEED.

THE term "cutting speed," as applied to machine tools, means the number of feet of cutting performed by the tool edge, in a given time, or what is the same thing, the number of feet the shaving, cut by the tool in a given time, would measure if extended in a straight line. The term "feed," as applied to a machine tool, means the thickness of the cut or shaving taken by the tool.

Planing machines being constructed so that their tables run at a given and unchangeable speed, their cutting speed is fixed; and the operator has only, therefore, to consider the question of the amount of feed to be given to the tool at a cut, which may be placed at a maximum by keeping the tool as stout as possible in proportion to its work, making it as hard as its strength will allow, and fastening it so that its cutting edge will be as close to the tool post as circumstances will permit. In all cases, however, cast-iron may be cut in a planer with a coarser feed than is possible with wrought-iron. Milling machines should have their cutters revolve so that the cutting speed of the largest diameter of the cutter does not exceed 18 feet per minute, at which speed the cut taken may be made, without injury to the cutter, as deep as the machine will drive.

It is only when we treat of lathe work that the questions of feed and speed assume their real importance, for there is no part of the turner's art in which so great a variation of practice exists or is possible, no part of his art so intricate and deceptive, and none requiring so much judgment, perception, and watchfulness, not only because the nature of

CUTTING SPEED AND FEED.

the work to be performed may render peculiar conditions of speed and feed necessary, but also because a tool may appear to the unpractised or even to the experienced eye, to be doing excellent duty, when it is really falling far short of the duty it is capable of performing. For all work which is so slight as to be very liable to spring from the force of the cut, for work to perform which a tool slight in body must be used, and in cases where the tool has to take out a sweep or round a corner which has a break in it, a light or fine feed must be employed; and it is therefore advisable to let the cutting speed be as fast as the tool will stand. But under all ordinary circumstances, a maximum of tool feed rather than of lathe speed will perform the greatest quantity of work in a given time. A keen tool, used with a quick speed and fine feed, will cut off a thin shaving with a rapidity very pleasing to the eye, but equally as deceptive to the judgment; for under such a high rate of cutting speed, the tool will not stand either a deep cut or a coarse feed; and the increase in the depth of cut and in the feed of the tool, obtainable by the employment of a slower lathe speed, more than compensates for the reduction of lathe speed necessary to their attainment, as the following remarks will disclose.

Wrought-iron, of about two inches in diameter, is not uncommonly turned with a tool feed of one inch of tool travel to 40 revolutions of the lathe. With a tool feed as fine as this, it is possible, on work of this size, to employ a cutting speed as high as 27 feet per minute, providing the depth of the cut does not exceed one-eighth of an inch, reducing the diameter of the work to $1\frac{3}{4}$ inches. The length of shaft or rod turned under such circumstances will be $1\frac{9}{32}$ inches per minute, since the lathe speed (necessary to give the tool a cutting speed of 27 feet per minute) would require to be about 51 revolutions per minute; and as each revolution of the lathe moves the tool forward $\frac{1}{40}$ of an inch, the duty performed is $\frac{51}{40}$ of an inch, or $1\frac{9}{32}$

inches of shaft turned per minute, as before stated. If, however, we turn the same rod or shaft of two inch iron, with a lathe speed of 36 revolutions per minute, and a tool travel of one inch to 24 revolutions of the lathe, the amount of duty performed will be $\frac{36}{24}$ inches, or $1\frac{1}{2}$ inches of shaft turned per minute. Here, then, we have a gain of about 17 per cent. in favor of the employment of the slow speed and quick feed. Nor is this all, for we have reduced the cutting speed to 19 feet, instead of 27 feet per minute, and the tool will, in consequence, stand the cut much longer and cut cleaner.

Pursuing our investigations still further, we find from actual test that, cutting at the rate of 27 feet per minute, the tool will not stand a cut deeper than one-eighth of an inch; whereas under the cutting speed of 19 feet per minute, it will take a cut of one-quarter of an inch in depth, thus considerably more than doubling the duty performed by the tool, in consequence of the decreased cutting speed and increased feed or tool travel.

Lathe work of about three-quarters of an inch in diameter may, if there is no break in the cut, be turned at a cutting speed of as much as 36 feet per minute, the feed being one inch of tool travel to about 25 revolutions of the lathe. The revolutions per minute of the lathe, necessary to give such a rate of cutting speed, will be about 183; the duty performed will therefore be $\frac{183}{25}$, or $7\frac{5}{16}$ inches of three-quarter inch iron turned per minute. A feed of one inch of tool travel to 25 revolutions of the lathe is greater than is generally employed upon work of so small a diameter as three-quarter inch, but is not too great for the generality of work of such a size; for the tool will stand either a roughing or smoothing cut at that speed, unless in the exceptional case of the work being so long as to cause it to spring away from the tool. Under these circumstances the feed may be reduced to one inch of tool travel to 30 or 40 revolutions of the lathe, according to the length and depth of the cut.

It will be observed that the cutting speed given, for work of three-quarter inch diameter, is nearly double that given as the most advantageous for work of two inches diameter, while the feed or tool travel is nearly the same in both cases; the reason of this is that the tool can be ground much keener for the smaller sized than it could be for the larger sized work, and, furthermore, because the metal, being cut off the smaller work, is not so well supported by the metal behind it as is the metal being cut off the larger work, and, in consequence, places less strain upon the tool point, as illustrated in Figs. 66 and 67.

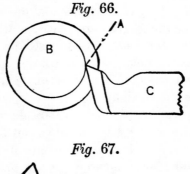

Fig. 66.

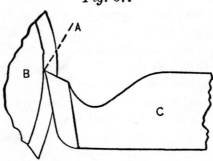

Fig. 67.

B is a shaft, and C is the tool in both cases. The dotted line a, in Fig. 66, does not, it will be observed, pass through so much of the metal of the shaft B, as does the dotted line a, of the shaft B, in Fig. 67. The metal in contact with the point of the tool in Fig. 66, is not, therefore, so well supported by the metal behind it as is the metal in contact with the point of the tool in Fig. 67, the result being that the tool, taking a cut on the smaller shaft equal in depth to that taken by the tool on the larger one, may have a higher rate of cutting speed without sustaining any more force from the cut, the difference in the resistance of the metal to the tools being equalized by the increased speed of the smaller shaft.

These conditions are reversed in the case of boring, the metal, being cut in a small hole, being better supported by the metal behind it than is the case in a larger hole or bore. This may be overcome by making the boring tool point cut below the horizontal centre of the work, while the body of the tool may, to keep it stout enough, be kept in the centre of the hole. But in a large bore, the effect is not so seriously encountered, because of the nearer approach of the circle to the straight line.

On heavy work it is specially desirable to have the tool stand a long time without being taken out to grind, for the following reasons: 1. It takes longer to stop and start the lathe, and to take out and replace the tool. 2. It takes longer to readjust the tool to its cut. 3. It takes more time to put the feed motion into gear again. 4. The feed motion is very slow to travel the tool up and into its cut, and to take up its play or lost motion. 5. Lastly, the tool should take a great many more feet of cut, at one grinding, than is the case with a tool for small work.

A tool used on work 5 inches diameter (the lathe making 20 revolutions to feed the tool one inch) would perform 314 feet of cutting in travelling a foot, the lathe having, of course, performed 240 revolutions; while one used on work 10 feet in diameter (with the same ratio of speed) will have performed 314 feet of cutting when the tool has travelled half an inch, and the lathe made 10 revolutions only. In practice, however, the feed for larger work is increased in a far greater ratio than the cutting speed is diminished, as compared with small work; but in all cases the old axiom and poetical couplet holds good, that

> "A quick feed
> And slow speed"

are the most expeditious for cutting off a quantity of metal, and in the case of cast-iron, for finishing it also.

A positive or constant rate of cutting speed for large work cannot be given, because the hardness of the metal, the liability of the work to spring in consequence of its shape, the distance of the point of the tool from the tool post, and other causes already explained, may render a deviation necessary, but the following are the approximate speeds and feeds for ordinary work:

TABLES OF CUTTING SPEEDS AND FEEDS.

Table for Steel.

Diameter of work in inches.	ROUGHING CUTS.		FINISHING CUTS.	
	Speed in feet per minute.	Feed.	Speed in feet per minute.	Feed.
1 and less	20	25	20	30
1 to 2	18	25	18	30
2 to 3	18	25	15	30
3 to 6	15	20	15	30

For Wrought-Iron.

Diameter of work in inches.	ROUGHING CUTS.		FINISHING CUTS.	
	Speed in feet per minute.	Feed.	Speed in feet per minute.	Feed.
1 and less	35	25	38	30
1 to 2	25	20	30	30
2 to 4	25	20	25	25
4 to 6	23	20	23	25
6 to 12	20	15	23	20
12 to 20	18	12	18	16

For Cast-Iron.

Diameter of work in inches.	ROUGHING CUTS.		FINISHING CUTS.	
	Speed in feet per minute.	Feed.	Speed in feet per minute.	Feed.
1 and less	38	20	38	20
1 to 2	35	20	35	16
2 to 4	30	20	30	10
4 to 6	25	16	25	6
6 to 12	20	14	20	6
12 to 20	20	10	20	4

For Brass.

Diameter of work in inches.	ROUGHING CUTS.		FINISHING CUTS.	
	Speed in feet per minute.	Feed.	Speed in feet per minute.	Feed.
1 and less	120	25	120	25
1 to 2	100	25	100	25
2 to 4	80	25	100	25
4 to 6	70	25	70	25
6 to 12	60	25	70	25

For Copper.

Diameter of work in inches.	ROUGHING CUTS.		FINISHING CUTS.	
	Speed in feet per minute.	Feed.	Speed in feet per minute.	Feed.
1 and less	350	25	400	25
2 to 5	250	25	300	25
5 to 12	200	25	200	25
12 to 20	150	25	150	30

In cases where the cuts are unusually long ones, the cutting speeds may be slightly reduced except in the case of copper. All the tools we have so far described may justly be termed master tools, for work on external surfaces, each entirely filling its arena, and all other tools used on outside work are simply modifications called into requisition to suit exceptional cases.

CHAPTER III.

BORING TOOLS FOR LATHE WORK.

STANDARD bits and reamers have superseded the use of boring tools for all special and many other purposes, but there are numerous cases where a boring tool cannot be dispensed with, especially in repairing shops and for promiscuous work.

Boring tools for use on lathe work require to be shaped with greater exactitude than any other lathe tools, for the reason that they are slighter in body in proportion to the duty required of them than any other; and as a rule, the cutting edges standing further out from the tool post or clamp, the body of the tool is more subject to spring from the strain of the cut. It is obvious that, if the hole to be bored out is a long one, the cutting edge of the tool will become dull at the end of the hole as compared to what it was at the commencement (a remark which, of course, applies to all tools); but in tools stout in proportion to the duty required of them, and held close in to the tool post, the effect of the slight wear of the cutting edge, due to a finishing cut, is not practically appreciable. In the case of a boring tool, however, the distance of the cutting edge from the tool post renders the slightest variation in the cutting capability of the tool sufficient to affect the work, as may be experienced by boring out a hole half of its length, and then merely exerting a pressure on the body of the tool, as near the entrance of the hole as possible, with the fingers, when the size of the last half of the hole will be found to have varied according to the direction in which the pressure was placed. As a result of this extreme sensitiveness to spring, the tool is apt to spring away from the cut as the boring proceeds, thus leaving the hole smaller at the back than at the front end. To remedy this defect, several very fine finishing cuts may be taken; but a better plan is to so shape the tool that its spring will be in a direction the least liable to affect the size of the bore of the work.

The pressure on the cutting edge of a tool acts in two directions, the one vertical, the other lateral. The downward pressure remains, under equal conditions, at all times the same; the lateral pressure varies according to the direc-

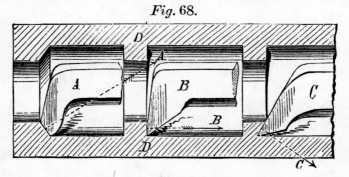

Fig. 68.

tion of the plane of the cutting edge of the tool to the line or direction in which the tool travels: the general direction of the pressure being at a right angle to the general direction of the plane of the cutting edge. For example, the lateral pressure, and hence the spring of the various tools, shown in Fig. 68, will be in each case in the direction denoted by the dotted lines. D is a section of a piece of metal requiring the three inside collars to be cut out; A, B, and C are variously shaped boring tools, from which it will be seen that A would leave the cut in proportion as it suffered from spring, which would increase as the tool edge became dull, and that the cut forms a wedge, tending to force the tool towards the centre of the work. B would neither spring it to nor away from the cut, but would simply require more power to feed it as the edge became dulled; while C would have a tendency to run into the cut in proportion as it springs; and as the tool edge became dull, it would force the tool point deeper and deeper into the cut until something gave way. Now, in addition to this consideration of spring, we have the relative keenness of the tools, it being obvious at a glance that (independent of any top rake or lip) C is the keenest, and A the least keen tool; and since wrought-iron requires the keenest, cast-iron a medium, and brass the least keen tool, it follows that we may accept, as a rule, C for wrought-iron, B for cast-iron, and A for brass work. To this rule there are, however, variations to be made to suit exceptional cases, such for instance as when a hole terminates in solid metal and has a flat bottom, in which case the tool B (slightly modified towards the form of tool C) must be employed. Or suppose a hole in cast-iron to be, as is often the case, very hard at and near the surface of the metal. Tool A would commence cutting the hard surface, and, becoming dull, would spring away from the cut in spite of all that could be done to prevent it; while tool B would commence to cut both the hard and the soft metal together, the cutting edge wearing rapidly away where it came into contact with the hard surface of the metal; and these conditions would, in both cases, continue during the whole operation of boring, rendering it difficult and tardy. But if the tool C were employed, the point of the tool would commence cutting the soft part of the metal first, and would undermine the hard surface, and (from the pressure) break it instead of cutting it away, as shown in Fig. 69, in which a is the point of the tool, and from a to B is the cutting edge; the dotted lines, c and D, represent the depth of the cut, c being the inside skin of the metal, supposed to be hard.

The angle at which the cutting edge stands to the cut causes the pressure, due to the bending and fracturing of the shaving, to be in the direction of *e*, which keeps the tool point into its cut; while the resistance of the tool point to this force, reacting upon the cut, from *a* to B, causes the hard skin to break away.

Fig. 69.

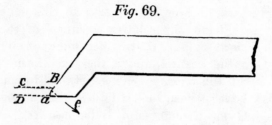

When a cut is being taken which is not sufficient to clean up or true the work, less top rake must be given, as a very keen tool loses its edge more quickly than one less keen. The reason for taking the rake off the top of a tool is that, if it were taken off the bottom, the cutting edge would not be so well supported by the metal, and would have a tendency to scrape, which rule applies both to inside and outside cuts. For brass work, top rake is never applied, because it would cause the tool to jar and cut roughly, bottom rake alone being sufficient to give a tool for brass the requisite keenness.

The application of top rake or lip to a boring tool lessens the strain due to serving the metal; by presenting a keener cutting edge, it lessens the tendency to lateral spring, and increases that to vertical spring, and is beneficial in all cases in which it can be employed. Upon wrought-iron and steel it is indispensable; upon cast it may be employed to a limited degree; and upon brass it is inadmissible by reason of its causing the tool to either jar or chatter. In Fig. 70, B represents a section of the work, No. 1 represents a boring tool with top rake, for wrought-iron, and No. 2 a tool without top rake, for brass work, which may be also used for cast-iron when the tool stands a long way out from the tool post or clamp, under which circumstances it is

BORING TOOLS FOR LATHE WORK. 75

liable to jar or chatter. A tool for use on wrought-iron should have the same amount of top rake, no matter how far it stands out from the tool post; whereas one for use on cast-iron or brass requires to be the less keen the further it stands out from the tool post. To take a very smooth cut on brass work, the top face of the tool, shown at 2 in Fig. 70, must be ground off, as denoted by the dotted line.

Fig. 70.

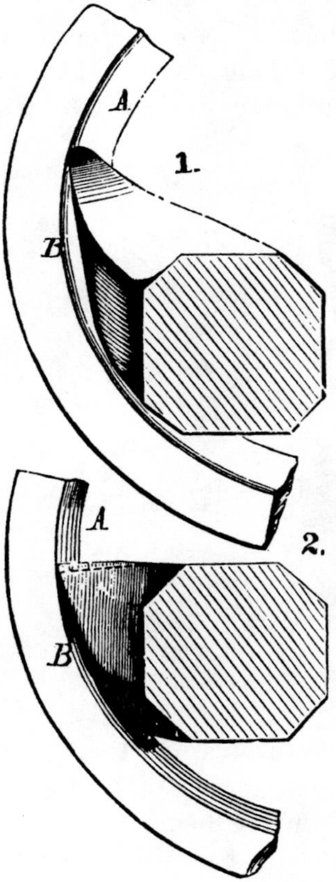

We have now to consider the most desirable shape for the corner of the cutting edge. A positively sharp corner, unless for a special purpose, is very undesirable, because the extreme point soon wears away, leaving the cutting qualification of the tool almost destroyed, and because it leaves the work rough, and can only be employed with a very fine feed. It may be accepted as a general rule that, for roughing cuts, on brass work, the corner should be sufficiently rounded to give strength to the tool point; while, in finishing cuts, the point may be made as round as possible without causing the tool to jar or chatter. Now, since the tendency of the tool to jar or chatter upon all metals depends upon four points, namely, the distance it stands out from the tool post, the amount of top rake, the acuteness or keenness of the general outline of the tool, and the shape of the cutting corner, it will readily be perceived that considerable judgment is required to determine the most desirable form for any particular conditions, and that it is only by understanding the principles governing the conditions that a tool to suit them may be at once formed.

In Fig. 71 will be found the various forms of boring tools for ordinary use. No. 1 is for use when the conditions admit of a heavy cut on wrought-iron. No. 2 is for use on wrought-iron when the tool stands so far from the tool post as to be necessarily subject to spring. No. 3 is to cut out a square corner at the bottom of a hole in wrought-iron. No. 4 is for taking out a heavy cut in cast-iron. No. 5 is

Fig. 71.

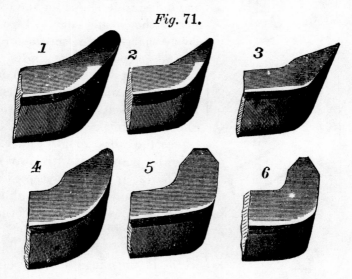

for taking out a finishing cut in cast-iron when the tool is proportionally stout, and hence not liable to spring or chatter: the point being flat, the cutting being performed by the front corner, and the back part being adjusted to merely scrape. No. 6 is for use on cast-iron under conditions in which the tool is liable to jar or spring.

An inspection of all these tools will disclose that the tool point is more rounded for favorable conditions, that is, when the body of the tool is stout, and the cutting edge is not held far out from the tool post; that, to prevent jarring, the

BORING TOOLS FOR LATHE WORK.

point of the tool is made less round, which is done to reduce the cutting surface of the tool edge (since it is apparent that, with a given depth of cut, the round-pointed tool will present the most cutting edge to the cut); and that, to further prevent jarring or chattering, the leading part of the cutting edge is ground at an angle; while, as another precaution against that evil, the general form of the tool is varied from that of tool C, in Fig. 68, towards that of tool A in the same figure; while for brass work, no top rake or lip is employed, but the tool is bevelled off to suit those cases in which it is liable to excessive spring. It is obvious that the feed may be coarser for a round-nosed than for a more acute tool, and that, the rounder the nose, the smoother the cut will be with the same rate of feed.

For heavy duty on wrought-iron, whether in large or small holes, the boring tool, represented in Fig. 72, has no

Fig. 72.

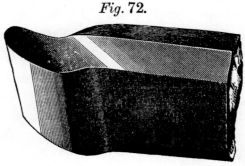

equal. The rake on the top face makes the cutting edge perform its duty on the front edge, and the strain due to bending the shaving tends to draw the tool to its cut, giving it an inclination to feed itself forward, thus relieving the feed screw of a part of the duty due to the strain of feeding.

The cutting edge should not stand above the horizontal level of the top of the tool body; otherwise, so stout a tool could not be gotten into a given size of hole; a consideration which, in small holes, is of the utmost importance. For similar duty on brass, the tool shown in Fig. 73 is the best that can be employed.

Fig. 73.

BORING TOOL FOR BRASS.

When, upon brass work, a boring tool has a broad cutting surface, such as is required to cut a recess, the only way to prevent extreme chattering and jarring is to grind off the top face, giving it negative top rake, as shown in Fig. 74: *a* being a section of the body of the tool, B the cutting part, and *c* the outline of the hole. B, being the lowest point of the top face, possesses negative top rake, and a corresponding tendency to scrape rather than cut keenly. The point B should always be above the centre of the hole, so that, in springing, it will spring away from and not into its cut. A boring tool, slight in proportion to its duty, and for use upon small wrought-iron work, should always be placed so that its cutting edge is a little below the centre of the hole, in which case the bottom of the body of the tool is liable, in small holes, to bear against the bottom of the hole, unless the cutting part is made to be a little below the centre of the body of the tool, rendering it rather difficult to grind on the top face. It is not, however, imperatively necessary to grind it there, since it can be sharpened by grinding the side faces;

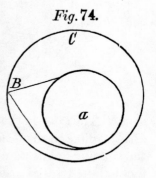

Fig. 74.

and the advantage gained by being enabled to get, into a given sized hole, a stouter tool than otherwise could be done, and, as a result, to take deeper and more nearly parallel cuts (for such tools generally spring off their cut at the back end of the hole, leaving it taper unless several light cuts are taken out), more than compensates for the extra wear of the tool, consequent upon being able to grind it upon one part only.

Boring tools for use on wrought-iron, cast-iron, steel or copper, require very little side or bottom rake, only sufficient, in fact, to well clear the sides of the cut, and the straighter these side faces are kept the stronger the tool; and the better the cutting edges are supported by the metal behind them, the longer will they stand without regrinding.

When boring light brass work, it is well to hold a brush near the entrance of the hole, to prevent the turnings from flying about the shop; while cutting tools for outside brass work may have a split-leather washer forced over the body near the cutting end for the same purpose.

After a piece of brass or cast-iron work has been bored and taken out of the lathe, and is found on trial to fit a little too tight, it may, if it is difficult to chuck it true again, be eased by a half-round scraper, as follows: Take an old half-round, smooth file, and grind the teeth completely out of the flat face; then grind the edges to an angle sufficiently acute to cut freely, as a scraper; then rechuck the work in the lathe as nearly true as possible, and revolve it at such a speed that the scraper will cut at about 380 feet per minute; then apply the scraper edge to the bore of the hole at the bottom, moving it along the bore and holding it firmly. If the flat face and the bevelled edge of the scraper be ground true and even, and care is taken in using it to take out the metal only where required, this tool will perform excellent duty and cut very smoothly. It may also be used to advantage to ease out by hand the narrow places of a hole that is oval, or the small end of one that is taper and requires to be made parallel. The smoothness of its work is much improved by smoothing its

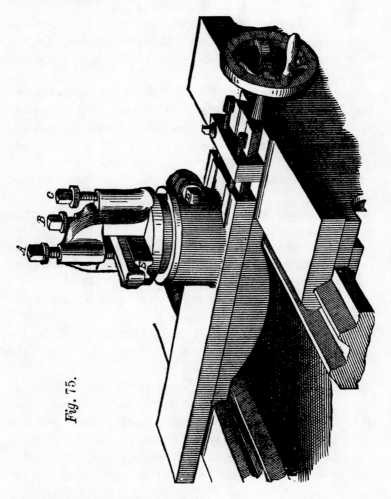

Fig. 75.

edge upon an oilstone. Here it may be well to state that the application of an oilstone to the cutting edges of a boring tool increases its tendency to chatter; if, therefore, a hole requires to be made unusually smooth, the tool

BORING TOOLS FOR LATHE WORK. 81

must be given less top rake and may then be oilstoned. In many cases a tool may be prevented from chattering by holding it with the fingers as near the entrance of the hole as possible.

Fig. 75 represents a boring tool, composed of a piece of octagon steel lying in a holder or seat E, which may be so set in the tool-post as to support the tool as near to the cutting edge as the depth of the hole to be bored will permit.

Fig. 76 represents a boring tool made of round steel, and clamped between two pieces 1 and 2, so that the tool

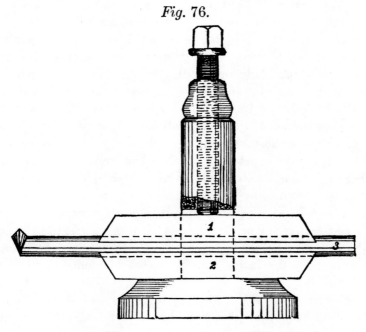

Fig. 76.

may be readily adjusted to the work by revolving it on its axis, and placed to project through the clamps as far as necessary for the work, and to facilitate this the tool is sometimes separate and secured by a set screw.

BORING TOOL HOLDERS.

For use on holes too small to admit of a bar having a sliding head, which are usually bored with a slide rest tool, a boring tool holder may be employed to great advantage. Such a holder may be made by squaring or flattening one end of a round bar of iron so that it will fit into the tool post of the lathe, and cutting into the opposite end a groove to receive a short boring tool, the latter being fastened to its place by set screws provided in the holder or by being wedged in with a small wedge. Various sizes of such holders should be made, the larger sizes being provided with set screws to hold the tool. For use in holes of from two to eight inches bore, such an appliance is invaluable, especially if the hole to be bored is of unusual depth; because the bar may be made very stout in proportion to the size of the hole, and will, therefore, stand a depth of cut and a rate of feed totally impracticable with an ordinary boring tool, and will not spring away from its cut towards the back end of the hole, as boring tools are apt to do. Furthermore, the cutting tools, being small, are easily forged, ground up, and renewed when worn out; and the bar maintains its original length, which may be made to suit the depth of hole required to be bored; while a boring tool becomes shorter each time it requires reforging.

For truing out broad recesses in large work, the slot in the end may be made large enough to receive two tools, one to turn the inside and the other the outside of the recess.

For use upon holes of a very large bore, or upon outside work in which the tool requires to stand a long way out from the slide rest, as sometimes occurs when the diameter of the work is so near the full size, the lathe will swing; that there is not sufficient room for the slide rest of the lathe to pass under the work, a square tool holder should be employed, such tool holder being a stout bar of square iron, say 2½ inches square, and having a complete tool box on one end, the tool box being provided with two stout steel set screws.

CHAPTER IV.

SCREW-CUTTING TOOLS.

LATHE tools for cutting screws have necessarily, from the nature of their duty, a comparatively broad cutting surface, rendering them very subject to spring. Those used for V threads, being ground to fit the V of the thread, are, in consequence, weak and liable to break; to avoid which they should only be given enough bottom rake to clear the thread well, and top rake sufficient to made them cut clean. They are used at a slow rate of cutting speed, and may therefore be lowered to a straw-colored temper (as reducing the temper strengthens a tool). Firmness and strength are of great importance to this class of tool, so that it should be fastened with the cutting edge as near to the tool post as is convenient.

For use on wrought-iron, the V or thread-cutting tool is sometimes given side rake; but this is not a necessity and is of doubtful utility, because the advantage gained by its tendency to assist in feeding itself is quite counterbalanced by its increased liability to break at the point. It should always be placed to cut at the centre of the work.

If the pitch of the screw to be cut is very coarse, a tool nearly one-half of the width of the space between one thread and the next should be employed, so as to avoid the spring which a tool of the full width would undergo. After taking several cuts, the tool must be moved laterally to the amount of its width, and cuts taken off as before until the tool has cut somewhat deeper than it did before being moved, when it must be placed back again in its first position, and the process repeated until the required depth of thread is attained.

Fig. 77.

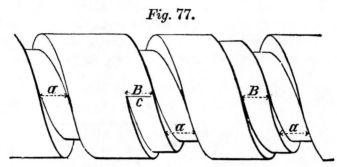

Fig. 77 represents a thread or screw during the above described process of cutting. *a a a* is the groove or space taken out by the cuts before the tool was moved; B B represents the first cut taken after it was moved; *c* is the point to which the cut B is supposed (for the purpose of this illustration) to have travelled.

The tool used having been a little less than one-half the proper width of the space of the thread, it becomes evident that the thread will be left with rather more than its proper thickness, which is done to allow finishing cuts to be taken upon its sides, for which purpose the side tool (given in Fig. 36) is brought into requisition, care being taken that it be placed true, so as to cut both sides of the thread of an equal angle to the centre line of the screw.

In cutting V threads of a coarse pitch, the tool may be made less in width than the required space between the threads demands, so that it may be moved a little laterally in order to take a cut off one side of the thread only at a time, by which means a heavier cut may be taken with less liability for the tool to spring in; but the finishing cut is better if taken by a tool of the full width or shape of the thread.

The most accurate method of cutting small V threads

is to use a stout chaser fastened in the tool post, and then feed it with the screw-cutting gear of the lathe, the same as with a common screw-cutting tool. Such a chaser should be made hollow in the length of the tooth, possess a minimum of top rake, and be placed to cut at the centre of the work; and it should be so placed in the tool post that the teeth stand exactly parallel to the line of the cut.

Fig 78.

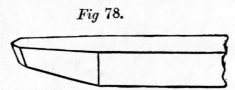

Fig. 78 represents a tool for cutting an outside V thread in brass work. When, however, the tool point must, of necessity, stand far out from the tool post, it must be given negative top rake, to make it cut smoothly and prevent its jarring. To adapt this tool to cutting V threads on iron, it is only necessary to give it top rake.

Fig. 79.

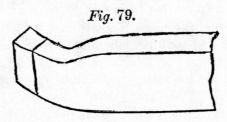

Fig. 79 represents a very stout tool, adapted to cutting coarse square threads on wrought-iron or steel. For cutting square threads on brass work, the tool shown in Fig. 80 should be used.

Fig. 80.

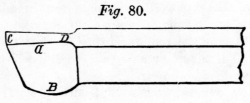

SCREW-CUTTING TOOLS.

Fig. 81 represents a single pointed tool, for cutting an internal thread, and it is obvious that in order that it may cut a thread of correct shape, it is necessary to grind the V to a gauge, to set it so that each cutting edge shall stand at an equal degree of angle to the work axis, and also that the cutting edges shall stand level with the centre of the work or line of lathe centres.

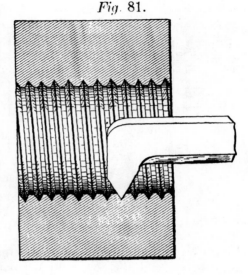

Fig. 81.

There are three different shapes of V threads in use in the United States, viz.: first, the

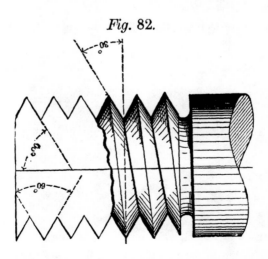

Fig. 82.

sharp V thread, shown in Fig. 82, in which the thread

sides meet in a point both at the top and bottom, and these sides are at an angle of 60 degrees to one another. Second, the United States standard, whose form is shown in Fig.

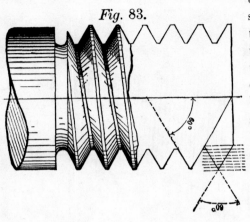

Fig. 83.

83, which corresponds in shape to the above thread, with one-eighth of the top cut off, and one-eighth of the bottom filled in so as to leave a flat place at the top and bottom of the thread. And third, the Whitworth thread shown in Fig. 84, in which the angle of the sides of the thread, one to the other, is 55 degrees, and the tops and bottoms of the thread are rounded.

Of these three threads, the sharp V or common V, as it is sometimes termed, is the easiest to produce, because

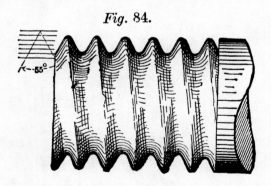

Fig. 84.

if the angles of the threading tool are correct the top and bottom will come correct of itself, whereas in the United States standard thread it is necessary to form the tool as in Fig. 85, the flat at the point requiring great care to make it of correct width. Either of these threads may, however, be originated more easily than the Whitworth or English standard thread, whose rounded tops and bottoms are very difficult to form correctly and exactly alike. On this account the Whitworth thread is usually cut by chasers cut from a standard hob, or master tap, the hob being revolved and the chaser pressed to it. When, however, a chaser is produced, it possesses the advantage that the angles of the thread sides are more correct than is the case with single-pointed tools ground upon a grinding stone. But the common V and the United States standard thread may both be cut by chasers, and in fact the United States standard thread may be cut by a chaser having the common V thread; all that is necessary being to grind off the tops of the teeth, as in Fig. 86, because, when the chaser

Fig. 85.

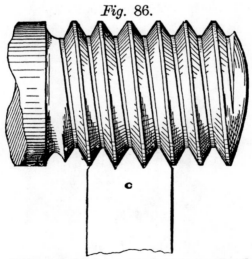

Fig. 86.

has entered the work far enough to cut the flat at the thread bottom to the correct diameter, the flat at the thread top will be left of itself.

8*

Fig. 87 represents a gauge for testing the angles of threading tools for the common V, and for the United States standard threads. It is not unusual, however, to

Fig. 87.

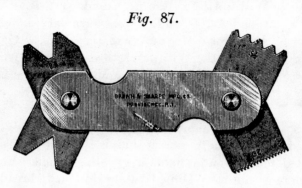

employ a short metal gauge, such as shown in Fig. 88, applying it direct to the work, which will test if the tool has been correctly set with relation to the work. But when it is known that the tool is correctly formed, and has been properly set in the lathe tool post and therefore

Fig. 88.

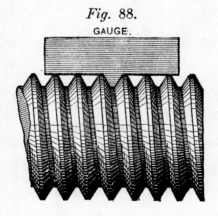

that the shape of the thread is correct, the thread diameter may be most correctly measured by callipers, applied as in Fig. 89, which is especially advantageous when there is at hand a correct thread to set the callipers by.

To test the pitch of a thread, to find if its pitch is alike at various parts of the thread length, gauges such as at

Fig. 89.

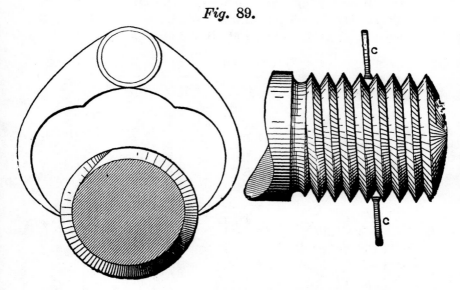

G or G′ in Fig. 90 may be employed; the work and the

Fig. 90.

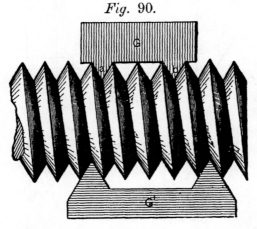

gauge being held up to the light, which will show very clearly any error that may exist.

92   COMPLETE PRACTICAL MACHINIST.

In cutting V threads upon either inside or outside work, great care should be taken to grind the V tool to the exact proper angle, and to also set it quite true in the lathe; to accomplish both of which results, we have the gauge shown in Fig. 91, and sold at all tool stores.

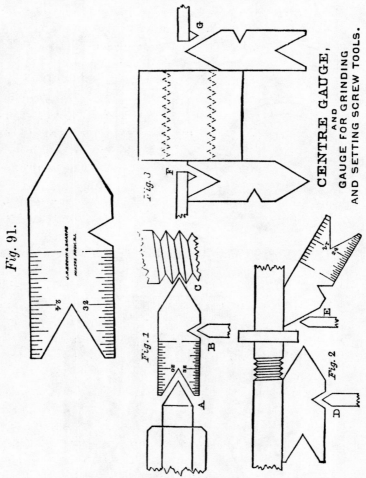

The above cuts show the various uses to which this gauge can be applied.

In Fig. 1, at A, is shown the manner of gauging the

SCREW-CUTTING TOOLS.

angle to which a lathe centre should be turned ; at B, the angle to which a screw-thread cutting tool should be ground, and at C, the correctness of the angle of a screw-thread already cut.

In Fig. 2, the shaft with a screw-thread is supposed to be held on the centre of a lathe. By applying the gauge as shown at D, or E, the thread tool can be set at right angles to the shaft, and then fastened in place by the screw in tool post, thereby avoiding imperfect or leaning threads.

In Fig. 3, the manner of setting the tool for cutting inside threads is illustrated. The angles used in this gauge are sixty degrees. The four divisions upon the gauge of 14, 20, 24 and 32 parts to the inch are very useful in measuring the number of threads to the inch of taps and screws.

The following parts to the inch can be determined by them—namely, 2, 3, 4, 5, 6, 7, 8, 10, 14, 16, 20, 24 and 32.

If the tool is not ground to the correct V, or is not set true in the lathe, the result is, that the threads will bear upon each other upon one side, or a portion of one side only—thus reducing the amount of wearing surface, and causing the threads to soon become a loose fit, as well as to be weaker than they should be. A V thread cut by a V tool in the lathe is not so strong as one cut by a chaser, because chasers cut a thread slightly rounded at the top and bottom; whereas the V tool leaves a sharp corner.

At the termination of the thread, it is necessary to cut a recess as deep as the thread, in order to give the chaser clearance, and prevent it from ripping into the shoulder, which would form the termination of the thread in the absence of a recess. It is a very common practice to cut this groove or recess with a V tool or graver point, instead of with a round-nosed tool, thus producing a recess having a conical instead of a curved outline: the result being to very seriously impair the strength of the bolt, and cause it, under severe strains, to fracture across the section of the bottom of the groove.

In a series of experiments made a few years ago, by the English government, upon targets representing ships' armor, the bolts were found to be unable to withstand the shock caused by the cannon shot striking the target; and it being observed that the fracture nearly always occurred across the section above referred to, the clearance grooves were made with a hollow curve, which obviated the defect.

To calculate the change gear wheels necessary to cut a given pitch of thread in a lathe:

The pitch of a thread is measured or denominated in two ways, first: by the number of threads there are in one inch of the length of the screw; and, second, by the distance of one thread from the next one. Thus, in Fig. 92,

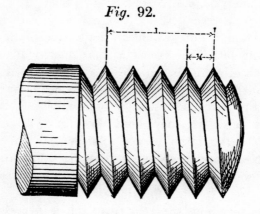

Fig. 92.

the thread may be expressed as one of four per inch, or as a thread, having a pitch of ¼ inch. For fine threads the pitch is usually given in the number of threads per inch. What is called (as applied to its screw-cutting wheels) a single geared lathe is one in which the driving gear is either fastened upon and revolves with the man-

SCREW-CUTTING TOOLS.

dril or spindle of the lathe, or else is driven by an intermediate gear-wheel of such a size that the driving gear, though not fast upon the lathe spindle or mandril, still makes the same number of revolutions per minute as does the mandril, while at the same time no two wheels (on such a lathe) of different diameters run side by side, making an equal number of revolutions in a given time.

Thus in Fig. 93 we have the driving gear D, and intermediate wheel I, and the lead screw wheel S, the

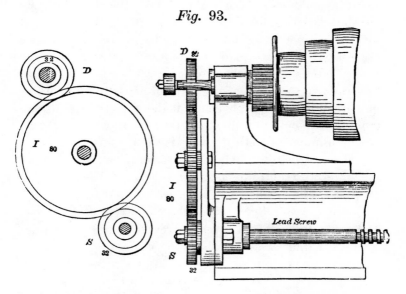

Fig. 93.

arrangement constituting a simple or single geared screw cutting lathe.

In such a lathe we have only to consider the driving wheel or gear and the gear upon the feed screw of the lathe, the others or intermediate wheels having no effect or influence (upon the thread to be cut) other than to make up the distance between the driving and feed screw gears, and thus to communicate the motion of the one to the other. Hence, having ascertained what sized wheel is required for the driving wheel and on the feed screw, we may connect them together by any wheel or wheels that will answer irrespective of their sizes.

It will be readily perceived, then, that if the driving gear and the feed screw gear contain respectively the same number of teeth, the lathe would be geared to cut a thread of the same pitch as the pitch of the thread on the feed screw of the lathe, because the feed screw would revolve at the same speed as the lathe did. Now, in exact proportion as the feed screw revolves slower than does the lathe spindle or mandril, will the thread cut by the lathe tool be finer than that on the feed screw, and *vice versa;* hence we have—whereby to find the wheels necessary to cut a thread of a required pitch in a single geared lathe—the following rule:

Put down the pitch of the lead screw, as the numerator of a fraction, and put beneath it the pitch of the thread you want to cut, and these figures will represent the required number of teeth the wheels should have.

For example, the pitch of a lead screw is 8 threads per inch, and we require to cut a pitch of 16. Then

$$\frac{\text{Pitch of lead screw } 8}{\text{Thread to be cut } 16} = \frac{\text{number of teeth for the driving gear.}}{\text{No. of teeth for the gear on the lead screw.}}$$

Again: the pitch of a lead screw is 6, and we require to cut a thread whose pitch is 24. Then

$$\frac{\text{Pitch of lead screw } 6}{\text{Pitch to be cut } 24} = \frac{\text{number of teeth for driving gear.}}{\text{number of teeth for lead screw gear.}}$$

If we have no wheels containing these respective numbers of teeth, we multiply the fraction by any number we may choose, as by 2, 3, 4, 5 or 6.

Thus—

$$\frac{6}{24} \times 2 = \frac{12}{48} = \frac{\text{number of teeth for driving gear.}}{\text{number of teeth for lead screw gear.}}$$

$$\text{Or—} \frac{6}{24} \times 3 = \frac{18}{72} = \frac{\text{number of teeth for driving gear.}}{\text{number of teeth for lead screw gear.}}$$

$$\text{Or—} \frac{6}{24} \times 4 = \frac{24}{96} = \frac{\text{number of teeth for driving gear.}}{\text{number of teeth for lead screw gear.}}$$

Let us now suppose that we require to cut a fractional thread, as, say, the lead screw being 4 per inch, we wish to cut a pitch of 4¼ threads per inch, and all we have to do is to put down the lead screw pitch expressed in quarter inches and the pitch to be cut in quarter inches. Thus there are 16 quarters in 4 and 17 quarters in 4¼. Hence, the fraction becomes

$\dfrac{16}{17}$ = number of teeth for driving gear.
= number of teeth for lead screw gear.

If we have not these gears, we multiply by any number as before. Thus—

$\dfrac{16}{17} \times 2 = \dfrac{32}{34}$ = number of teeth for driving gear.
= number of teeth for lead screw gear.

Or— $\dfrac{16}{17} \times 3 = \dfrac{48}{51}$ = number of teeth for driving gear.
= number of teeth for lead screw gear.

Or— $\dfrac{16}{17} \times 4 = \dfrac{64}{68}$ = number of teeth for driving gear.
= number of teeth for lead screw gear.

The term *Compound* or *double geared*, as applied to the screw-cutting gear of a lathe, means that there exists, between the gear wheel which is fastened to, and revolves with, the lathe spindle, and the feed screw, two gear wheels of different diameters and revolving side by side, at the same number of revolutions, by reason of being fixed upon the same sleeve or axis. The object of this arrangement is to make, between the speed at which the lathe mandril or spindle will run, and the speed or revolutions at which the feed screw will run, a greater amount of difference than is possible in a single geared lathe, and thus to be able to cut threads of a coarser pitch than could be cut in the latter. This is usually accomplished by providing two intermediate wheels of different diameters, both being held by a feather to a sleeve revolving upon an adjustable pin provided for the purpose. Thus Fig. 94 represents an arrangement of compounded gear, in which A is the driving gear, C and D the compounded pair of wheels carried on a stud in the swing frame F, and S the lead screw gear. In this arrangement, the driving gear A is fixed and cannot readily be taken off; hence it must in many cases be taken into account in finding the gears, all the changes being made in the wheels C, D and S.

Sometimes, however, only the wheel upon the lead screw need be changed, since a wide range of pitch may be obtained without disturbing any other wheel. Suppose, for example, that the driving gear or gear on the lathe mandril or spindle has 32 teeth, and that the compounded pair is arranged to reduce the motion one-half, and the effect is the same as if the driving gear had 16 teeth and there was no compound gears employed.

Fig. 94.

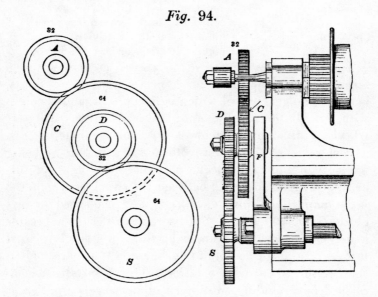

To find the number of teeth in the wheel required to be placed on the feed screw, we have the following rule:

Divide the pitch to be cut by the pitch of the feed screw, and the product will be the proportional number. Then multiply the number of teeth on the lathe mandril gear by the number of teeth on the smallest gear of the com-

SCREW-CUTTING.

pounded pair, and the product by the proportional number, and divide the last product by the number of teeth in the largest wheel of the compounded pair, and the product is the number of teeth for the wheel on the feed screw.

Suppose, for example, the gear on the lathe mandril contains 40 teeth running into the largest of the compounded gears which contains 50 teeth, and that the small gear of the compounded pair contains 15 teeth; what wheel will be required for the feed screw—its pitch being 2, and the thread requiring to be cut being 20?

$$\underset{\text{Pitch required.}}{20} \div \underset{\text{Pitch of feed screw.}}{2} = \underset{\text{Proportional number.}}{10}$$

Then—

$$\underset{\text{Mandril gear teeth.}}{40} \times \underset{\text{Small compound gear.}}{15} \times \underset{\text{Proportional number.}}{10} \div \underset{\text{Large compound gear.}}{50} = 120 =$$ the number of teeth required upon the wheel for the feed screw. In the above example, however, all the necessary wheels except one are given; and since it is often required to find the necessary sizes of two of the wheels, the following rule may be used:

Divide the number of threads you wish to cut by the pitch of the feed screw, and multiply the quotient by the number of teeth on one of the driving wheels, and the product by the number of teeth on the other of the driving wheels; then any divisor that leaves no remainder to the last product is the number of teeth for one of the wheels driven, and the quotient is the number of teeth for the other wheel driven.

[In this rule the term "wheel driven" means a wheel which has motion imparted to it, while its teeth do not drive or revolve any other wheel; hence the large wheel of the compounded pair is one of the wheels driven, while the wheel on the feed screw is the other of the wheels driven.]

Example. It is required to cut 20 threads to the inch, the pitch of the feed screw being 2, one of the driving wheels contains 40 teeth and the other 15:

100 COMPLETE PRACTICAL MACHINIST.

Pitch required to be cut.		Pitch of feed screw.		Teeth in one driving wheel.		Teeth in other driving wheel.		
20	÷	2	×	40	×	15	=	6000.

Then, $6000 \div 50 = 120$; and hence one of the gears will require to contain 50 and the other 120 teeth; if we have not two of such wheels, we may divide by some other number instead of 50.

Thus: $6000 \div 60 = 100$; and the wheels will require to have, respectively, 60 and 100 teeth.

If there are no wheels on the lathe we proceed as follows:

Divide the pitch required by the pitch of the feed screw; the quotient is the proportion between the revolutions of the first driving gear and the feed screw gear.

Example. Required the gears to cut a pitch of 20, the feed screw pitch being 4; here $20 \div 4 = 5$; that is to say, the feed screw must revolve five times as slowly as the first driving gear; we now find two numbers which, multiplied together, make five: as $2\frac{1}{2} \times 2 = 5$; hence one pair of wheels must be geared $2\frac{1}{2}$ to 1 and the other pair 2 to 1, the small wheel of each pair being used as drivers, because the thread required is finer than the feed screw.

Fig. 95 represents an arrangement of compound gears

Fig. 95.

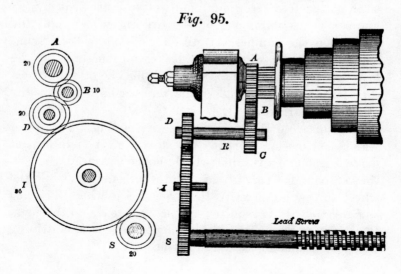

SCREW-CUTTING.

common in small American lathes. A is the actual driving gear, B an intermediate, and C D are the compound pair. In this case the wheels A, B and C are fixed and

Fig. 96.

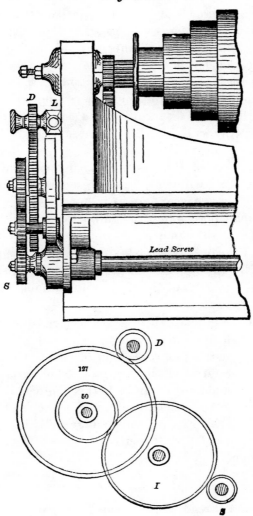

cannot be changed; hence all we have to consider is the sizes of D, and of the lead screw gear S. Suppose, now, that the wheel D has the same number of teeth as wheel C, and we may neglect C and calculate the change gears the same as if D was on the lathe spindle, and A, B and C were not used. But in a majority of cases D will have to be changed as well as S, and then the size of C must be taken into account. In this case we proceed precisely as before, finding the proportion that must exist between the revolution of the mandril and of the lead screw, and arranging the wheels accordingly.

The wheels necessary to cut a left-hand thread are obviously the same as those necessary for a right-hand one of the same pitch.

The pitches of threads used in France are given in terms of the centimeter, and the method of finding the necessary change gears are as follows:

An inch equals $\frac{254}{100}$ of a centimeter, or in other words 1 inch bears the same proportion to a centimeter as 254 does to 100, and we may take the fraction $\frac{254}{100}$ and reduce it by any number that will divide both terms of the fraction without leaving a remainder. Thus $\frac{254}{100} \div 2 = \frac{127}{50}$.

If then we take a pair of gears, having respectively 127 and 50 teeth, they will form a compound pair that will enable the cutting of threads in terms of the centimeter instead of in terms of the inch. It is obvious that as a centimeter is more than an inch, this compound pair must be used to reduce the revolutions of the lead screw, or arranged as in Fig. 96, the changes of wheels for any given number of threads per centimeter being made at D and at S only.

TO CUT A DOUBLE THREAD.

A double thread is one formed by two spiral grooves instead of one. Thus in Fig. 97 we have one spiral at A and another at B, the latter being carried as far as C only. The true pitch of the thread is in this case the pitch of one spiral, or twice the apparent pitch.

To cut such a thread, we arrange the change wheels for the true pitch A and cut that spiral first. Then we stop the lathe and taking off the lead screw gear of the change wheels, and move the lathe so that the driving gear makes one-half revolution; then we put the lead screw gear back and the lathe is adjusted to cut the second thread.

Suppose, for example, that the wheels used are a 36

Fig. 97.

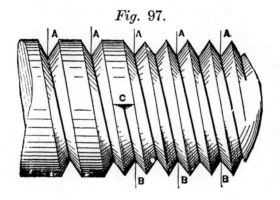

and a 72, as in Fig. 98; then we make a mark at S on the driving gear, and a corresponding one on the lead screw gear, and then take the lathe off the lead screw; we then count 18 teeth on the driving gear, make a mark on the eighteenth tooth and pull the lathe round so that the mark on the lead screw gear will engage with the eighteenth tooth on the lead screw gear.

For a treble screw we would require to divide the driving gear by three, thus, $36 \div 3 = 12$, and we must count 12 teeth from S and proceed as before; for a quadruple thread we divide the 36 by 4 and proceed as before, and so on.

Fig. 98.

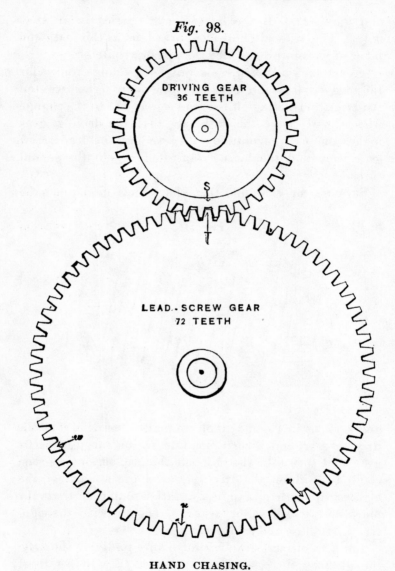

HAND CHASING.

To cut a screw by hand in the lathe we proceed as follows: The work is turned up to the required size, and then on the outside of the work we employ the V tool

SCREW-CUTTING TOOLS.

shown in Fig. 99; which tool is made of a piece of steel about $\frac{3}{16}$ or $\frac{1}{4}$ inch thick, and $\frac{5}{8}$ inch deep, the holding end being fitted into a handle. The point A is the cutting edge; the point B being formed so that when the tool is pressed firmly to the lathe rest face, it will not slip but will hold fast; and the top face being given a little top rake when the tool is used upon wrought-iron or steel, whereas negative top rake is necessary for use upon brass work.

The end of the work from which the thread starts should be filed smooth, and all the turning tool marks effaced before the attempt is made to start the thread, because the slightest obstruction will cause the motion of the starting tool to be irregular, and this will prevent the chaser from

Fig. 99.

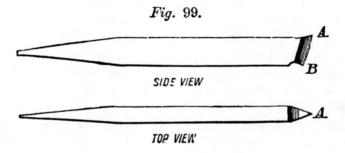

SIDE VIEW

TOP VIEW

readily picking up the thread, which is a delicate operation, requiring great care, even from an experienced hand.

When the thread starts from the end of the work, it is necessary to round off the corner, because (assuming the thread to be a right hand one) it is easier to start the thread at the right hand end, and carry it forward with the chaser, than it is to start it at the left hand end and carry it back to the right. Similarly for a left hand thread, it must be started at the left hand end, and carried to the right, and in this case the corner may be rounded off, either before or after chasing the thread at pleasure.

To start the thread, the lathe should be run at a fast speed; and the heel of the tool being pressed firmly to the face of the lathe-rest, the point of the V of the tool being brought firmly into contact with the work, while the handle of the tool must be twisted from right to left at the same time as it is moved bodily from the left to the right. It is the relative quickness with which these combined movements are performed which will determine the pitch of the thread. The results of these combined movements will be a fine groove cut upon the work, and of the same distance from one groove to the next as the distance of one tooth of the chaser to the next. If the spiral groove so cut is only the proper pitch at one part, as, say, at the starting end, the chaser may be so held and applied as only to touch that end, when it will readily find the groove if applied lightly to it. Then several light cuts may be taken off that end, before attempting to carry the thread along.

The chaser is applied by being pressed lightly against the work, and moved along the lathe-rest at as nearly the proper speed as can be judged. The chaser should be held so that its hind teeth press hardest against the work, which will keep them in the starting groove and act as a guide to the front teeth, while they extend the groove, carrying the thread forward to the required distance on the work.

The reason for running the lathe at a comparatively fast speed is, that the tool is then less likely to be checked in its movement by a seam or hard place in the metal of the bolt, and that, even if the metal is soft and uniform in its texture, it is easier to move the tool at a regular speed than it would be if the lathe ran comparatively slowly.

If the tool is moved irregularly or becomes checked in its forward movement, the thread will become "drunken," that is, it will not move forward at a uniform speed; and if the thread is drunken when it is started, the chaser will not only fail to rectify it, but, if the drunken part occurs in a part of the iron either harder or softer than the rest of the metal, the thread will become more drunken as the chaser proceeds. It is preferable, therefore, if the thread is not started truly, to try again, and, if there is not sufficient metal to permit of the starting groove first struck being turned out, to make another further along the bolt. It takes much time and patience to learn to strike the requisite pitch at the first trial; and it is therefore requisite for a beginner to leave the end of the work larger in diameter than the required finished size, so as to have metal sufficient to turn out the first few starting grooves, should they not be true or of the correct pitch. If, however, a correct starting groove is struck at the first attempt, the chaser may be applied sufficiently to cut the thread down to and along the body of the bolt; then the projection may be turned down with the graver to the required size, and the chasing proceeded with.

After the thread is struck, and before the chaser is applied to it, the top face of the rest should be lightly filed to remove any burrs which may have been made by the heel of the V tool or graver; or such burrs, by checking the even movement of the chaser, will cause it to make the thread drunken. Where the length of the thread terminates, a hollow curved groove should be cut, its depth being even with the bottom of the thread; the object of this groove is to give the chaser clearance, and to enable you to cut the thread parallel from end to end and not to leave the last thread or two larger in diameter than the rest. Another object is to prevent the front tooth of the chaser from ripping in and breaking off, as it would be very apt to do in the absence of the groove.

TO MAKE A CHASER.

Chasers are cut from a hub, that is to say, a cutter formed by cutting a thread upon a piece of round steel, and then forming cutting edges by cutting a series of grooves along the length of the hub. These grooves should be V-shaped, the cutting side of the groove having its face pointing radially towards the centre of the hub. Hubs should be tempered to a brown color. A chaser is made from a piece of flat steel whose width and thickness increases with the pitch of the thread; the following proportions will, however, be found correct:

NUMBER OF THREADS PER INCH.	NUMBER OF TEETH IN THE CHASER.	THICKNESS OF THE CHASER.
24 to 20	12 to 14	$\frac{1}{4}$ inch.
18 to 14	10	$\frac{5}{16}$ "
12 to 8	9 to 6	$\frac{5}{16}$ "
6 to 4	7 to 6	$\frac{3}{8}$ "

The end face of the chaser should be filed level and at an angle with both the top face and the front edge of the steel, both top and bottom rounded off so that at the top it will not dig into the shoulder at the end of the thread, and at the bottom it will not strike against a burr or other obstruction or the face of the lathe rest, and thus be retarded in its forward movement while being cut. The hub is then driven in the lathe between the centres, the chaser being held in a handle sufficiently long to enable the operator to hold it with one hand, and press the shoulder against the end so as to force the end of the chaser against the hub, which will of itself carry the chaser along the rest. During the operation of cutting the chaser by the hub, the former will be upside down, its cutting face (when finished) being that which during this operation is resting on the face of the lathe-rest, which latter should be placed a short distance from, and not close up to the hub. After the chaser has passed once down the hub, special attention should be paid as to whether the front tooth will become a full one; if not, the marks cut by the hub should be filed out again, and a new trial essayed. It must be borne in mind that, the chaser being held upside down, the back tooth, while cutting the chaser, becomes the front

SCREW-CUTTING TOOLS.

one when the chaser is reversed and ready for use. The hub should be run at a comparatively slow speed, and kept freely supplied with oil, it being an expensive tool to make, and this method of using preserves it. In Fig. 100, A is a chaser whose front tooth is not a full one; B is a chaser with a full front tooth; and it is obvious that the half tooth at A would break.

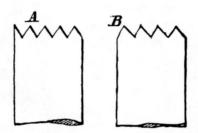

Fig. 100.

The cutting operation of the hub upon the chaser is continued until the thread upon the latter is cut full, when it is taken to the vise and filed as follows:

The top and bottom edges immediately behind the front tooth are rounded off as already directed. Then along the bottom of the chaser the teeth are rounded off to prevent them from catching against any burr on the face of the lathe rest.

An outside chaser for cutting wrought-iron by hand should be made hollow in the length of the tooth, and have top rake, to enable it to cut easily; for the strain required to bend the shaving out of the straight line will hold the teeth to their cut. Top rake may, in fact, be applied to such an extent that the chaser will cut well of itself without having any force applied to it except sufficient to keep it level, but if made so keen, it soon loses its edge and is very apt to break.

For use on cast-iron or brass, an outside chaser must be

made less keen by giving the top face of the teeth no rake, or else negative top rake and cutting the teeth less hollow in their lengths. The latter object is obtained by moving the handle, in which the chaser is fixed, up and down while the hub is cutting it.

The lathe rest should be so adjusted that the chaser teeth cut above the horizontal centre of the work. The teeth of the chaser should fit the thread on the bolt along all their length when the body of the chaser is horizontal, and then the least raising of the handle end of the chaser will present the teeth to the work in position to cut, while the teeth behind the cutting edge will fit the thread being cut, sufficiently close to form a guide to steady the chaser. This method of using will not only keep the thread true, but will preserve the cutting edge of the chaser. If a chaser has top rake, and the handle end is held too high and so that the back of the teeth are clear of the thread, it will cut a thread deeper than are its own teeth; if, on the other hand, the top face is beveled off, and the handle is held too high, it will cut a thread shallower than the chaser teeth.

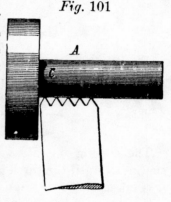

Fig. 101

Fig. 101 represents a chaser in use on wrought-iron. It will be observed that the tops of the teeth do not stand at a right angle to the side edges of the chaser; the object of this is to make the front edge of the chaser clear the driver or dog driving the work.

An inside chaser, that is, one for cutting threads in a hole or bore, should be, if to be used for cutting a right-handed thread, cut off a left-handed hub, otherwise the chaser will have its thread sloping in the opposite direction to the thread to be cut. This is shown as follows:

SCREW-CUTTING TOOLS.

In Fig. 102 is shown a top view of an inside chaser applied to a piece of work represented to be cut in half so as to expose the chaser to view. Now in order to en-

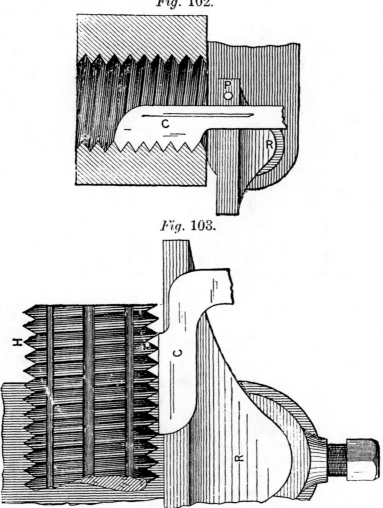

Fig. 102.

Fig. 103.

able the cutting of the chaser from the hob it must be bent as shown in Fig. 103, in which C represents the chaser, R the lathe hand rest, and H the hob.

An end view is shown in Fig. 104, and it is seen that the chaser teeth will slant in the direction of the dotted

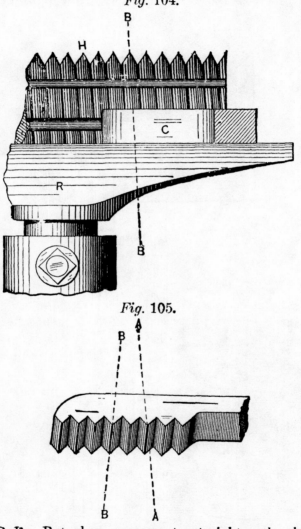

Fig. 104.

Fig. 105.

line B B. But when we come to straighten the chaser and turn it around as we must, to apply it to the work, we shall find that the teeth slant in the wrong direction, as is shown in Fig. 105, in which the dotted line B B corresponds to the same line in Fig. 104; whereas the teeth should slant in the direction of the dotted line A A, in order to match the threads in the work.

An inside chaser cut from a hob having a right-hand thread can be used to cut a right-hand one, but only by so tilting the teeth that only their edges have contact with the bore of the work. Now since an inside chaser would be too keen and would hence rip into the work if it possessed any top rake, and since it usually requires to have a slight degree of negative top rake, tilting it causes it to cut a thread shallower than the depth of its own teeth.

In the absence of a hob an inside chaser may be cut by a piece of wrought-iron having a hole and a slot cut in the side of the hole. If then the chaser is forged straight ready for use and fastened into the slot, and the hole is tapped out, the tap will at the same time cut the teeth upon the chaser, a right-hand tap cutting a right-hand chaser. In adopting this plan, however, it is proper to use a tap of a diameter large in proportion to the pitch of the thread, otherwise the teeth of the chaser will be hollowed too much in the direction of their length, and will in consequence jar or chatter when cutting, especially when in use upon long bores, in which cases the teeth cut at a long distance out from the lathe rest. It is a good plan to bore a quarter-inch hole in the top face of the lathe rest, and to insert therein a small pin, against which the edge of the chaser opposite to the teeth may be pressed, so that the pin will act as a fulcrum to force the teeth into their cut.

Inside threads are started by pressing the teeth lightly against the bore of the work, and moving the chaser forward at about the requisite speed. The corner of the bore of the work should be slightly rounded off (as should also the corner on the end of work to be chased with an outside thread) to prevent the chaser teeth from catching against it.

Either an inside or an outside chaser may be employed to cut a double or even a triple thread. A double thread is one in which the distance from one thread to the next is only one-half the actual pitch of the thread. Thus supposing a thread of five to an inch to be started in a screw-cutting lathe, and that the tool point is then moved laterally so as to cut another groove between the grooves first cut, there will be two threads each of a pitch of five to an inch, and yet the distance from one thread to the next will only be one-tenth of an inch, hence a chaser of the latter pitch may be used to cut up the two threads, thus producing a double thread whose actual is twice that of its apparent pitch.

Beginners should always stop the lathe and examine a single inside thread as soon as it is struck, for it is an easy matter to cut a double female thread in consequence of moving the chaser too fast, nor will the error be discovered until the thread is finished.

Double outside or male threads, to be cut by hand, can be most easily started by the chaser, moving it twice as fast as would be required for a single thread, rounding off the corner of the bolt end, and taking care to cut principally with the hindermost teeth.

The proper temper for the teeth is a deep brown, or, for unusually hard metal, a straw color. For chasing wrought-iron, the lathe may be run so that the teeth will perform about 40 feet, for steel about 30 feet, for cast-iron 50 feet, and for brass about 80 feet, of cutting per minute.

The quickest way to cut a number of threads upon bolts requiring to have a true thread and of an ordinarily good fit, is to take about two good cuts with a screw tool in the lathe, and then fastening a solid die in the vise to screw the bolt through the solid die by the aid of a wrench on the bolt head. The cuts taken in the lathe will make the bolt enter the die easily and true, while the die will insure correctness of size in the thread; bolts threaded thus may be screwed at least four times as quick as by finishing them entirely in the lathe.

In making a hob or master-tap for use to tap solid dies, cut in it as many flutes as will leave sufficient strength to the teeth, and let the number be an odd one.

To clean rusted threads on studs in their places, or to remove burrs from them, make a steel nut and file two slots through it after the manner of a solid die; and, after tempering it to a light straw color, screw it along the threads requiring to be cleaned, applying a little oil. It must not be forgotten, that as steel shrinks in hardening, the tap used for this purpose should be a little above the standard size, or else worked sufficiently in the nut to cut it out larger than the normal size.

CHAPTER V.

LATHE DOGS, CARRIERS OR DRIVERS.

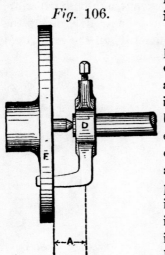

Fig. 106.

The simplest form of carrier or dog for lathe work is the bent-tailed dog, shown in Fig. 106, its bent end projecting into a slot in the face plate. It is objectionable, however, inasmuch as it is driven at the leverage A from the work it exerts a strain tending to bend it. This may be to some extent obviated by leaving its end straight and driving it with a pin projecting from the face plate, as in Fig. 107. The driving strain may be further equalized by employing two pins, as in Fig. 108, but it is difficult to bring both the driving pins to bear upon the dog. This may, however, be accomplished by the means shown in Fig. 109, which represents a face plate with the two pins thread into nuts in a T groove provided on the face plate. The pins are screwed up moderately tight upon the work and the cut is put on. If one pin only meets the dog it will slip in the groove and cause the other pin also to drive, and both pins may then be screwed firmly home to their nuts.

A more perfect method of equalizing the driving strain is by means of the Clements driver, which is self-adjusting

LATHE DOGS, CARRIERS OR DRIVERS.

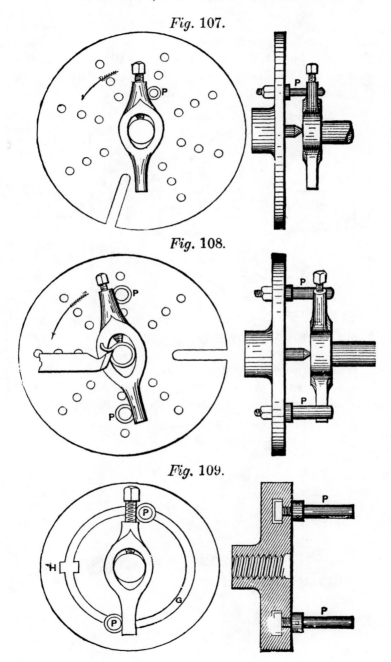

Fig. 107.

Fig. 108.

Fig. 109.

and is constructed as shown in Fig. 110. The plate F has four slots as A B, and through these and the face plate pass bolts C D, on which are small sliding blocks fitting into the slots in F. The work driving pins P P are threaded into nuts that are in T grooves, pro-

Fig. 110.

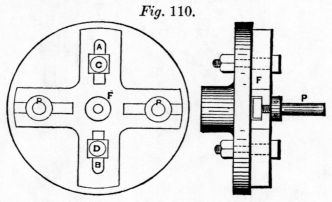

vided in F. When the pins meet the work driver the plate F moves upon the face plate, giving both pins an equal degree of driving pressure.

When the work requires to be driven backwards as well as forward, as in the case of screw-cutting, the dog may be secured by a set screw, such as E in Fig. 111.

Fig. 111.

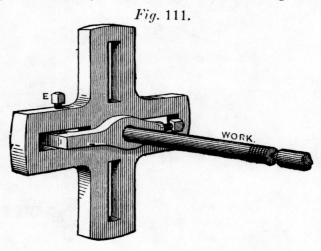

LATHE DOGS, CARRIERS OR DRIVERS. 119

For taper threads, however, the set screw must allow the dog end to move in the slot.

For driving bolts the driver may be formed as in Fig. 112 and bolted to the face plate, which saves the trouble

Fig. 112.

of fastening the driver to each bolt. Fig. 113, which is taken from "The American Machinist," represents an adjustable driver of this kind. One of the jaws, it will be observed, fits into a dove-tailed slide-way, and a screw is provided whereby the width of opening between the jaws may be adjusted to suit the size of the work. Drivers of this kind are especially suitable for small work, as they project less and are therefore less in the way than ordinary drivers, and may also be made thinner so as to accommodate thin bolt heads.

Fig. 113.

Fig. 114 represents the wood turner's **spur** centre, the wings being straight on the outside and coned within, so as to compress the wood around the central point and thus keep it true while at the same time obviating the liability to split the work.

Fig. 114.

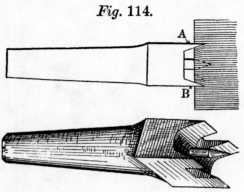

For short work the wood worker uses the screw chuck shown in Fig. 115, the work being centred and driven by the conical screw. For work that is true, the face A

Fig. 115.

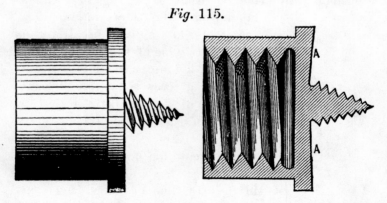

A may be made hollow as shown, which tends to true the work. Mandrils or arbors, as the smaller sizes are usually termed, should have their centres formed as in Fig. 116, the countersink being double, or else there should be a flat recess turned about the countersink, the object being

LATHE ARBORS.

in both cases to prevent the blows given to drive the mandril into the work from bruising the centres and causing them to run out of true. Mandrils should be made

Fig. 116.

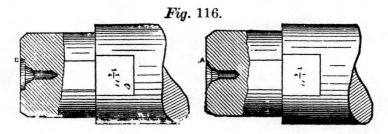

slightly taper, and made of wrought-iron, or what is better, hardened steel.

Fig. 117 represents an adjustable arbor or mandril

Fig. 117.

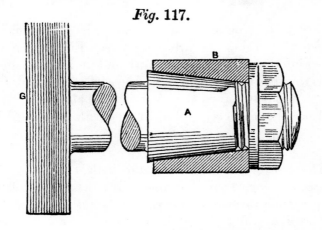

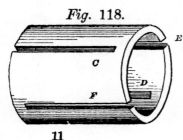

Fig. 118.

The body A is coned and the sleeve is split, as shown in Fig. 118, so that by means of the nut the sleeve may be forced up the arbor and its diameter made to unscrew to fit the bore of the work. Expanding mandrils are especially useful for holes that are reamed, because as the reamer wears, the size of hole it produces diminishes and will not fit a solid standard parallel arbor.

Fig. 119 represents a threaded arbor for work that is tapped, and it is seen that if the hole is not tapped quite true with the face the work will cant over and the facing

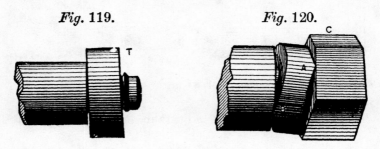

Fig. 119. *Fig.* 120.

will not be true with the thread axis. This may be avoided by using the arbor shown in Fig. 120, a ring being interposed between the work and the arbor shoulder. This ring has two diametrically opposite projections, A on one side and two B on the other, which balances the work and permits it to become locked true with the thread.

CENTRING LATHE WORK.

The centres of a lathe should both be of the same degree of cone so that the work will not wear to different shapes when turned end for end in the lathe.

The live centre should be tempered to a blue, which will preserve it, while leaving it quite soft enough to enable it to be turned up to true it with a fully hardened cutting tool. If a centre grinding device is at hand the centre may also be hardened to a straw color. Fig. 121 represents a centre grinding device for attachment to the lathe slide-rest in connection with the tool post. It consists simply of a countershaft above, driven from a pulley on the lathe live spindle.

The countershaft drives an emery wheel spindle below, which has end motion through its bearings, so that it may be fed to and fro along the cone of the lathe centre. The

Fig. 121.

emery wheel spindle is set at such an angle that the wheel operates a rear side of the lathe centre, so that the wheel and the centre revolve in opposite directions.

The quickest method of centring lathe work is by

means of a centring machine, such as in Fig. 122, which consists of a live spindle to drive the drill, and countershaft and a universal chuck to hold the work which should be freely supplied with oil during the drilling process.

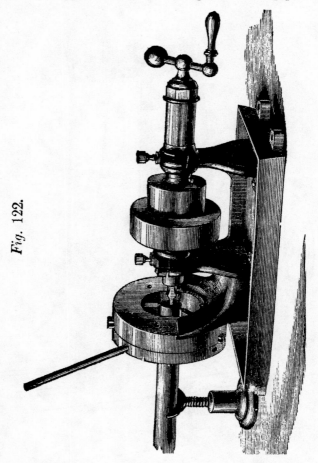

Fig. 122.

Fig. 123 represents a centre-drilling attachment for lathe work. In the tail spindle T is a cup or coned chuck D to hold that end of the work W true. S is a standard bolted to the lathe shears and carrying two fixed pins P which are each enveloped by a spiral spring. G is a piece having arms fitting over the pins P, and is capped or covered to receive the other end of the work. A small hole through the centre of G admits the drill. It is obvious that when the tail spindle T is fed up, it will feed the work to the drill, the piece G moving with the work against the pressure of the springs or pins P.

Fig. 123.

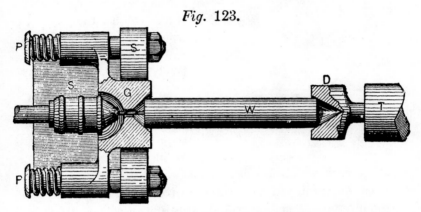

Fig. 124 represents a combined drill and countersink for centre drilling, the drill and countersink being in one piece. Fig. 125 represents a combined drill and counter-

Fig. 124. *Fig. 125.*

sink, in which a small twist-drill is let into the countersink and secured by a small set screw S so that the drill may be moved outward as it wears shorter. When very true work is required it is preferable to so shape the countersink that the lathe centre will first bear at the smallest part of the cone as is shown in Fig. 126. This will cause the countersink to wear and keep true with the hole.

If the centre drilling is to be done by hand it is very

11.*

important to relax every few seconds the hold upon the work sufficiently to permit it to make about a third of a revolution, which may be done while the other hand is supplying oil to the drill. The object and effect of this is to cause the centre drilling to be true, which otherwise it would not be, especially if the work is comparatively heavy, or heavier on one side than on another.

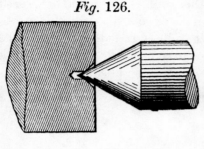

Fig. 126.

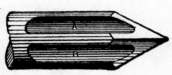

Fig. 127.

If, however, the work requires to run very true, as in the case of recentring work which has once been turned, the square centre must be employed to cut the centre of the work true to the circumference. A square centre is a centre fitted to the lathe in the same manner as the common centre, but having four flat sides ground upon its conical point, all four sides meeting at the point, and having sharp edges as shown in Fig. 127, the flutes serving to reduce the area of surface to be ground up when sharpening the cutting edges.

To recentre work that has already been turned, the square centre is put in the tailstock spindle of the lathe, in the same way as the ordinary centre is placed, the work having a dog or driver placed on it, as if the intention were to take a cut with the work placed in the lathe between the centres. A piece of iron or steel, having a hollow or flat end (as, for instance, the butt end of a tool) must then be fastened in the tool post of the lathe; then the lathe may be started and the tool end wound against the end of the work (close to the square centre)

CENTRING LATHE WORK.

until it touches it and forces it to run truly, in which position the tool end is left, while the square centre is fed up and into the work until the latter is true, when the operation will be completed. Before any turning is done to the diameter of any lathe work which runs between the centres, the ends of such work should be made true; because if there be a projecting part on the end, or if the latter is not quite true, the centre gradually moves over to the lowest side, as shown in Fig. 128, it being obvious that the countersink would move over as it wore from the side C towards the side D of the work.

All work which requires to be turned at both ends (and hence must be turned or placed end for end in the lathe) should be roughed out (that is, cut down to nearly the required size) all over before any part of it is finished, or, when turned end for end in the lathe, the part first turned up will run out of true with the part last turned up, though

Fig. 128.

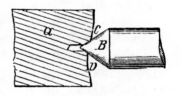

the lathe centres may be correctly placed. This may be caused by the centres of the work moving a little as they come to their bearings on the lathe centres, or in consequence of breaking the skin of the work; for nearly all work alters in form as its outside skin is removed, especially work in cast-iron.

FINISHING LATHE WORK.

The process to be adopted in finishing lathe work depends upon the degree of polish it is required to have.

Small work may be given the highest degree of polish by the use of the file and emery paper. The finishing cut should be taken with a sharp tool and as smoothly as possible, so as to have as little work as posssible for the file to do, because much filing will make the work out of round. Nothing coarser than a dead smooth file should be used, the work being run at a quick speed, or say at a circumferential speed of not less than about 170 feet per minute.

The file should be applied lightly to the work and with quick strokes, for if the file is held stationary, the filings will become locked in the file teeth, forming pins, which will cut scratches.

To prevent this we may apply either chalk or oil to the file and clean it after one dozen strokes or so, so as not to permit it to clog. When chalk is used, simply brushing the hand over the file will suffice.

EMERY CLOTH AND PAPER.

For ordinary work the common grades of emery paper and cloth may be employed, the finest being flour emery cloth or paper. The same grade of emery will cut coarser if placed on cloth than if on paper, because the surface of the cloth is not so smooth and even as that of the paper, and the consequence is that the grains of emery which are attached to the high spots on the cloth present a keener cutting edge and surface to the work than the rest of the surface. The main advantage of emery cloth lies in that it will wear longer because it is not so apt to tear. To fit emery cloth or paper for very fine work it should be used upon the work until the entire surface becomes worn even and glazed; the more it is worn and glazed the finer it will finish, and this remark applies equally to all kinds of emery cloth and paper, or crocus cloth. There is, however, an emery paper much finer than any other, its grades ranging from 1 to 0000, and it will produce a finish so fine as to give the work a finish and appearance equal to the finest silver or nickel-plating.

The method of using to produce a really fine finish is to revolve the work very fast in the lathe and to keep the

EMERY POLISHING.

emery paper moving rapidly, endwise of the work, so that the marks shall cross each other at a very obtuse angle. The coarser grades of cloth should be applied first, each successive grade being used until it has entirely removed the marks left by the grade previously used. The final polish is given by number 0000 paper, moved laterally

Fig. 129

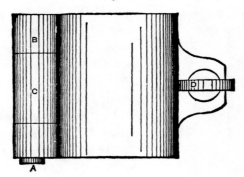

Fig. 130.

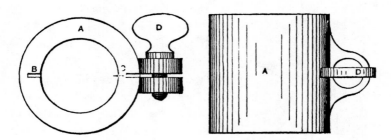

along the work very slowly, and under a very light pressure. To prepare the paper for the final finishing we must take the 0000 paper, and, giving the work a coating of oil barely sufficient to dull the polish, apply the paper, continually reversing its position in the hand so that all parts will become worn, the effects of the slight oiling being to cause the particles of metal cut off the work to adhere to and form a glaze upon the surface of the emery paper, and all metals polish best by being rubbed with a glazed surface composed of minute particles of the same metal as themselves; it follows, then, that the more emery paper or cloth becomes worn the finer it will polish.

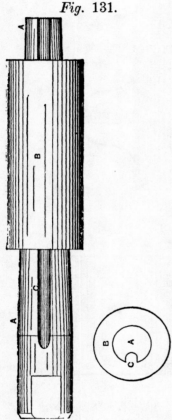

Fig. 131.

For larger and rougher work the filing may be done by an ordinary smooth file and succeeded by a polishing clamp consisting of two pieces of soft wood hinged by leather and containing holes to receive different sizes of work. The work is supplied with grain emery and oil, and is run at a quick speed, the clamp being closed firmly upon it and gradually moved to and fro. Towards the last of the process no fresh emery is applied, which makes the polish more perfect.

Grinding clamps for finishing work to gauge diameters are made as in Fig. 129, the two hinged halves being made of cast-iron recessed so as to receive a lining of babbitt metal, and held together by a screw D and pin A. In the small sizes a split bush, such as in Fig. 130, will serve.

For grinding out bores an arbor, A A, Fig. 131, is employed, having in it a groove C. B is a babbitt metal bush cast on its projection into the groove C serving as a driving key. As the diameter of B decreases, it may be driven further up the arbor or mandril, which, being taper, will expand with B.

LATHE CHUCKS.

Lathe chucks may be divided into three classes, as follows:

1st. Those in which the jaws are actuated simultaneously, which are called universal chucks.

2d. Those in which the jaws are actuated separately, which are called independent chucks, and

3d. Those in which the mechanism is so devised that the jaws may be operated either separately or independently at will, which are termed combination chucks.

Fig. 132.

Fig. 133.

Figs. 132 and 133 represent the Horton two-jawed chuck, with false or slip faces which are removable, so that jaws, having gripping surfaces of various shapes to suit the shape of the work, may be employed. The slips are dovetailed into the jaws and further secured by pins.

Fig. 134 represents what is called a box-body chuck, which is used to hold the brass turner's work. In

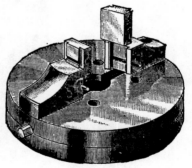

some of these chucks the jaws are operated simultaneously by a right and left-hand screw, while in others, each jaw has its own separate screw.

In the larger sizes of chucks, there are usually either three or four jaws. In a three-jawed chuck the work will be held with equal pressure by each jaw, because the fulcrum of the bite of each jaw is taken off the other two jaws, while in a four-jawed chuck, two opposite jaws may take all the strain, leaving the other two free from contact with the work. It is obvious, therefore, that for rough work, or work

Fig. 134.

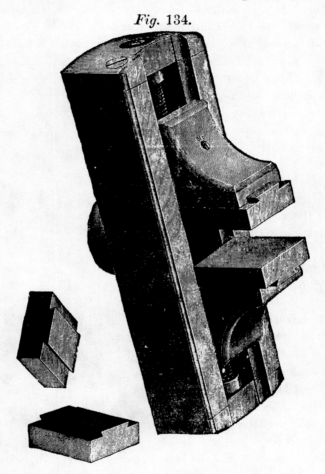

that is not cylindrical, three jaws are preferable to four, but if the work is true, then the four jaws are preferable, inasmuch as they hold the work at four points instead of at three.

CHUCKS.

Fig. 135 represents the Sweetland Chuck, which may be used as an independent or as a universal chuck. Each of the screws for operating the jaws is provided with a bevel pinion, and behind these pinions is a ring provided with teeth, and which may be caused to engage with or disengage from the pinions as follows: The width of the rack has a beveled step, the outer being thicker than the inner diameter. Between this ring or rack and the face of the chuck is placed, beneath each jaw, a cam block

Fig. 135.

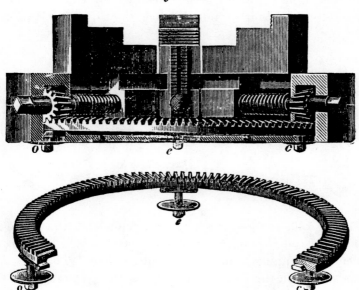

beveled to correspond with the beveled edge of the ring step.

Each cam block stem passes through radial slots in the face of the chuck, so that it may be moved towards or away from the centre of the chuck. When it is moved in, its cam-head passes into the recess or thin part of the ring-rack which then falls back out of gear with the jaw-screw pinion. But when it is moved outward the cam-head slides (on account of the beveled edge) under the ring-rack and places it in gear with the jaw-screw pinion. Thus to change the chuck from an independent one to a universal one all that is necessary is to push outwards the bolt-head of the cam-block stems, said heads being outside the chuck. The washers beneath these heads are dished to give them elasticity and enable them to steady the cams without undue friction.

To enable the jaws to be set true for using the chuck as a universal one, a circle is marked on the chuck face, and to this circle the edges of all the jaws must be set before operating the cams to put the rack ring in gear.

Fig. 136.

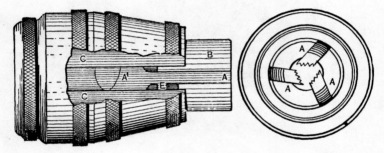

Fig. 136, represents a new drill chuck by the Russell Tool Co., of Boston, Mass., the object of which is to prevent the slipping common to small chucks. The jaws A are placed in the line of strain so as to drive rather than pull the work, and are serrated to increase the grip. The piece B moves out with the jaws to support them, and the jaws are provided with lugs E, which afford them extra support. As a result of these features, the chuck will hold sufficiently firmly to permit of its being used to drive work (having a diameter equal to the full capacity of the chuck) to be turned with the lathe tools.

CHUCKS.

Chuck dogs are detached dogs which fit into the square holes of the chuck plate or face plate, being held to the plate by a nut and washer. These dogs are movable to any part of the plate, their position being regulated to conform to the shape of the work, which renders possible their employment in cases where a dog chuck would be of no service; such, for instance, as holding a triangular or irregular shaped piece of work. The centre line of the screw should stand exactly parallel to the face of the face plate, or tightening the screws, which in this case grip the work, will force the latter towards or away from the face of the plate, according to the direction in which the screws are out of true. The screws should have their ends turned down below the thread, and should be hardened as directed for bell chuck screws, since these screws may be also reversed in the dog for some kinds of work. The dog should be screwed very firmly against the face plate, so as to avoid their springing.

Universal or scroll chucks, containing screws or gear wheels which are enclosed, should be occasionally very freely supplied with oil, and the chuck worked so as to move the jaws back and forth to the extreme end of their movement, so as to wash out any particles of metal or dust which may have lodged or collected in them; for proper cleaning will reduce the natural wear to a minimum, and prevent the internal parts from cutting, as they are otherwise apt to do.

When the work is liable to spring, from the pressure of the jaws of a chuck, those jaws may be slacked back a little previous to taking the finishing cut, during which the work need not be held so tightly.

From what has been already said it will be obvious that it is of great importance that, in addition to the jaws of a chuck being well fitted to the plate, there should be a large amount of wearing surface, so as to prevent as far as possible the jaws wearing loose in their slides.

CHAPTER VI.

TURNING ECCENTRICS.

If an eccentric has a hub or boss on one side only of its bore (as in the case of those for engines having link motions, where it is desirable to keep the eccentrics as close together as possible in order to avoid offset either in the bodies or double eyes of the eccentric rods), the first operation to be performed in turning it up is to chuck it with the hub side towards the face plate of the lathe, setting it true with its outside diameter (irrespective of the hole and hub running out of true), and to then face up the outside face. It must next be chucked so that the face already turned will be clamped against the face plate, setting the eccentric true to bore the hole out, and clamping balance weights on the face plate, opposite to the overhanging part of the eccentric. The hole, the face of the hub, the hub itself (if it is circular), and the face of the eccentric must be roughed out before any of them are finished, when the whole of them may be finished, to the requisite sizes and thicknesses. The eccentric must then be turned about and held to the chuck-plate by a plate or plates clamping the hub or boss only, the diameter of the eccentric being set true to the lines marked to set it by; then the diameter of the eccentric may be turned to fit the strap, the latter having been taken apart for that purpose. The reason for turning the strap before the eccentric is turned is (as may be inferred by the above) that the strap can be fitted to the eccentric while the latter is in the lathe, whereas the eccentric cannot be got into the strap while the strap is in the lathe. By this method, the outside of the eccentric will be turned true with a face that has been turned at the same chucking at which the hole was bored; while the eccentric will stand sufficiently far from the chuck to permit of the strap being tried on when it is necessary. And, moreover, the skin of the metal will have been removed on three out of the four faces before either of the working parts (the bore and the outside diameter) is finished; and as a consequence, the work will remain true, and not warp in consequence of the removal of the skin. Furthermore, upon the truth of the last chucking only will the truth of the whole job depend; and if the face plate of the lathe is a trifle out of true, the eccentric will only be out to an equal amount. It is not an uncommon practice (but a very reprehensible one) to face off the plain side of the eccentric, and to then bore the hole and turn the outside diameter, with the plain face clamped in both cases to the face plate. The fallacy of this method lies in the fact that, by such a procedure, the eccentric will be, when finished, out of true to twice the amount that the face plate is out of true.

.The strap should have a piece of thin sheet tin placed between the joint of the two halves before it is turned out, which tin should be taken out when the turning is completed, and the strap bolted together again. The size for the eccentric will then be from crown to crown of each half of the strap.

The object of inserting the tin is to make each half of the eccentric bed well upon the crown, and to prevent it from bearing too hard upon the points, as all straps do if the joint is not kept a little apart during the boring process. If the eccentric is already turned, an allowance may be made for the thickness of the sheet tin between the strap joint by placing a piece of the same tin beneath one of the caliper points when gauging the eccentric to take the size for the strap.

Eccentrics having a proportionally large amount of throw upon them are sometimes difficult to hold firmly, while their outside diameters are being turned to fit the strap, because the hub which is bolted against the face plate is so far from the centre of the work that, when the tool is cutting on the side of the eccentric opposite to the hub, the force of the cut is at a considerable leverage to the plates clamping the eccentrics; and the latter are, in consequence, very apt to move if a heavy cut is taken by the tool. Such an eccentric, however, usually has open spaces in its throw, which spaces are placed there to lighten it; the method of chucking may, under such circumstances, be varied as follows: The outside diameter of the eccentric may be gripped by the dog chuck, if the dogs of the chuck project far enough out to reach it (otherwise the dogs may grip the hub of the eccentric), while the hole is bored and the plain face of the eccentric turned. The eccentric must then be reversed in the lathe, and the hub and the face on that side must be turned. Then the plain face of the eccentric must be bolted to the face plate by plates placed across the spaces which are made to lighten the eccentric, and by a plate across the face of the hub. The eccentric being set true to the lines may then be turned on its outside diameter to fit the strap; to facilitate which fitting, thin parallel strips may be placed between the face plate and the plain face of the eccentric at this last chucking. It will be observed that, in either method of chucking, the outside diameter of the eccentric (that is to say, the part on which the strap fits) is turned with the face which was turned at the same chucking at which the hole was bored, clamped to the face plate. In cases where a number of eccentrics having the same size of bore and the same amount of throw are turned, there may be fitted to the face plate of the lathe a disk of sufficient diameter to fit the hole of the eccentric, said disk being fastened to the face plate at the required distance from the centre of the lathe to give the necessary amount of throw to the eccentric. The best method of fastening such a disk to the face plate is to provide it with a plain pin turned true with the disk, and let it fit a hole (bored in the face plate to receive it) sufficiently tightly to be just able to be taken in and out by the hand, the pin being provided with a screw at the end so that it can be screwed tight, by a nut, to the

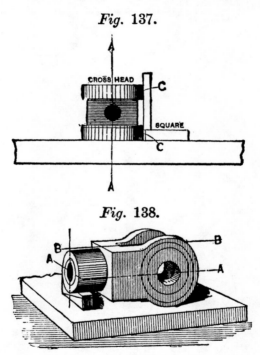

Fig. 137.

Fig. 138.

face plate. The last chucking of the eccentric is then performed by placing the hole of the eccentric on the disk, which will insure the correctness of the throw without the aid of any lines on the eccentric which may be set as true as the diameter of the casting will permit, and then turned to fit the strap. A similar disk, used in the same manner, may be employed on cranks, to insure exactness in their throw.

TURNING CRANKS.

A crank having a plain surface on its back should have such surface planed true. The large hole should be bored

Fig. 139.

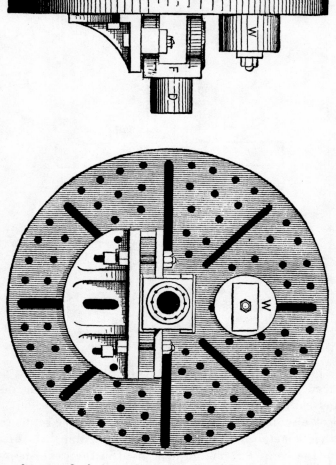

first, the crank being clamped with its planed surface to the chuck plate of the lathe, when the hole may be bored and the face of the hub trued up. To bore the hole for the crank pin, clamp the face of the hub of the crank, which has been trued up, against the plate of the lathe (the crank pin end of the crank being as it were suspended); then bolt two plates to the chuck plate, one on each side of the crank at the end to be bored, and place them so that their ends just come in contact with the crank end.

TO CHUCK A CROSSHEAD.

The bores of a crosshead must be at a right angle with the axes intersecting, and to accomplish this great care is necessary in marking the lines that are to be used in

Fig. 140. *Fig.* 141.

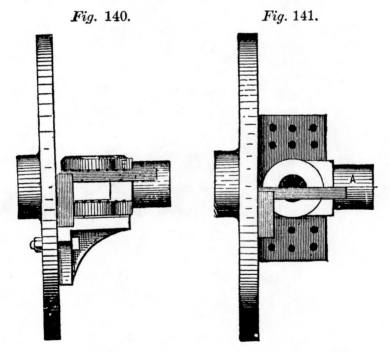

chucking it in the lathe. When the forging is true enough, it is the best plan to let the crosshead cheeks rest upon the marking-off table or plate, without any paper or other packing beneath them, as shown in Fig. 137, so that the square may be set against these two edges in the subsequent chuckings. If the edges are out of true, so that the work will not be true if marked out by them, they

Fig. 142.

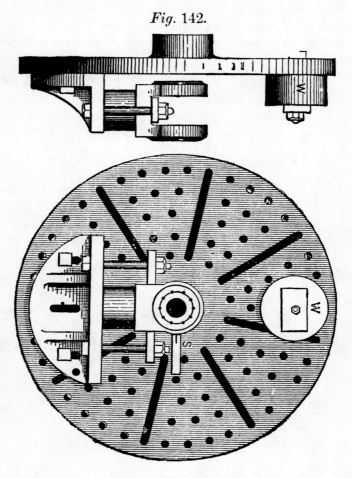

should be made true. This being done, the crosshead should be laid upon a plate, as in Fig 138, and the centre line A A marked around it. The centre line B B is next to be marked at a right angle to A A, and to do this the crosshead should be turned over on the table, as in Fig. 137, and squared by the edges C C of the cheeks, the line A A standing vertical. When B B is drawn, we may mark off the holes from the intersection of A A with B B, and the thickness of the cheeks from line B B, and the crosshead is ready to chuck.

The first chucking should be as in Fig. 139, one cheek being laid upon and bolted to an angle plate, the crosshead being set to the dotted circle on the end face of the hub and tested with a square applied to the dotted line F in Fig. 140, and also to the centre line A in Fig. 141, so that the crosshead may be set square, as well as having the circle run true. At this chucking the hole for the piston-rod would be bored, and the hub would be faced and turned. The second chucking would be as in Fig. 142, the faced end of the hub being bolted to the angle plate, and a square being applied to the edges C C, as in Fig. 143, while the dotted face of the circle on the face of the hub is set to run true, when the cheeks may be bored and faced inside and out, with the assurance that the work will be true and the holes at right angles to one another.

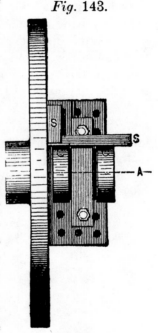

Fig. 143.

COUNTERBALANCING WORK.

When work is to be counterbalanced, the weight should be such as will effect the counterbalance when placed at a distance from the line of centres equal to the distance from that line of the heaviest part to be counterbalanced, and when the counterbalancing is to be done on work held between the centres, for example in the case of crank shafts, it is preferable to bolt the weight to the work itself and not to the faceplate of the lathe. In the absence of proper counterbalancing the work is apt to be turned elliptical.

BORING LINKS OR LEVERS.

In boring a number of lever arms or other work having holes requiring to be of precisely the same distance apart, we bore and finish one with great exactitude. Then after that one is bored, and the faces of the hub are faced off true with the hole, a pin, as shown in Fig. 144, should be made, the diameter of the part A being made to neatly fit one of the holes in the end of the arms or levers, and being made longer in length than is the length of the lever hole into which it fits. B is a washer, turned to fit easily to the diameter of A, and C is a collar, solid with A. D is a stem, turned parallel and true; and it is a little

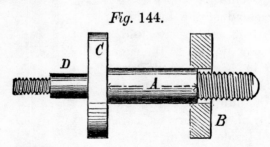

Fig. 144.

less in length than the thickness of the chuck plate upon which the arm is to be held while the holes are being bored. Upon each end a screw is provided to receive a nut. The use of this stud is as follows: Upon the chuck plate of the lathe or boring machine, and at the requisite distance from the centre, is bored a hole to receive at a close fit the plain part D, of the stud; and into this hole that end of the stud is fastened by means of a nut. One

TURNING LEVERS.

end of the lever or arm (being bored to fit the part A of the stud) is placed thereon, the stud being bolted to the chuck plate while the hole at the opposite end is being bored: thus insuring that the holes are exactly the same distance apart in all the levers. The manner of chucking is shown in Fig. 145 in which A represents a portion of the chuck, B the lever or arm to be bored, C the stud, and D D the plates bolted against the chuck so that their ends contact with the stem of the work to prevent it from

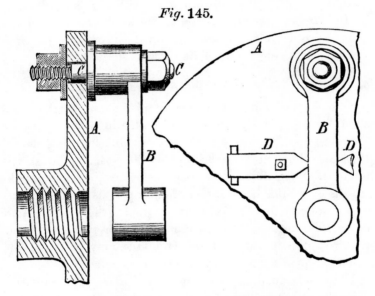

Fig. 145.

moving sideways during the operation of boring. The use of this stud, modified in shape to suit the work, is also applied to the turning of cranks, eccentrics, and other similar work, requiring unusual exactitude in the position of a hole or holes, or of a diameter in its position relative to a hole.

TURNING PISTONS AND RODS.

A piston should first be bored to receive the piston rod. The next operation is to rough out the body of the piston rod and to then fit it to the piston. The piston is then made fast to the rod, by the key, the nut, or by riveting, as the case may be, and the piston and rod should then be turned between the centres. By this means, the piston is sure to be true with the rod, which would not be the case if the piston and rod were turned separately. In turning the piston follower, that is, the disk which bolts to the piston head to retain the rings in their places, slack back the dogs or jaws of the chuck after the roughing out is complete, taking the finishing cuts with the jaws clamped as lightly as possible upon the work; because when the jaws of a chuck are screwed upon the work with great force, they spring it out of its natural shape.

PISTON RINGS.

The rings of metal from which piston rings are turned should have feet cast upon one end, which feet must be faced up true by taking a cut over them. The ring should then be chucked by bolting the faced feet against the chuck plate, so that the ring shall not be sprung in chucking, as it would be if it were held upon its inside or outside diameter by the jaws of a chuck. The inside and outside diameters of the ring may then be turned to their required dimensions, and the end face may be trued up, when the piston rings may be cut off as follows:

First introduce the parting tool, leaving the ring sufficiently wide to allow of a finishing cut after cutting the ring nearly off; introduce a side tool, shown in Fig. 36, and take a light finishing cut off the side of the ring, and then cut it off. The end face of the ring in the lathe may then be trued up by a finishing cut being taken over it, when the parting tool may be introduced and the process repeated for the next ring.

Piston rings are sometimes made thick on one side and thin on the other side of the diameter, the split of the ring being afterwards cut at its thinnest part, so that, when the ring is sprung into the cylinder (which is done to make the ring fit the cylinder tight and to cause it to expand as it wears, thus compensating for the wear), its spring will be equal all over and not mainly on the part of the diameter at right angles to the split, as it otherwise would be.

The process of turning such rings is to face the feet of the ring from which they are to be cut, and then turn up the outside diameter to its required size. Then move the ring on the face plate sufficiently to cause it to revolve eccentrically to the amount of the required difference between the thickest and thinnest parts of the ring, when the inside diameter should be trued out, and the rings cut off as before directed.

The object of turning the inside bore after and not before the outside diameter of the ring is turned, is that, during the process of cutting off the individual piston rings, the bore of the ring will be true, so that the parting tool will not come through the ring at one side sooner than at the other; for if this were the case, the parting tool, from its liability to spring and its broad cutting surface (parallel to the diameter of its cut), would be apt to spring in, rendering the cutting off process very difficult to perform; because if the piston ring is cut completely through on one side and not on the other, it will probably bend and spring from the pressure of the parting tool, and in most cases break off before being cut through at all parts by the tool.

The inside diameter (or bore) of piston rings is frequently left rough, that is to say, not turned out at all; but whenever this is the case, the splitting of the ring will in all probability cause one end of the ring (where it is split) to move laterally one way and the other end to move the opposite way, causing the vise hand a great deal of labor to file and scrape the sides of the ring true again. The cause of this spring is that there is a tension on the inside of the ring (where it has not been bored), tending to twist it, which tendency is overcome by the strength of the ring so long as it is solid; but when it is split, the tension releases itself by twisting the ring as stated.

The tension referred to is, in all probability, caused, to a certain extent, by the unequal cooling of the ring after it is cast.

Iron and brass moulders generally extract castings from the mould as soon as they are cool enough to permit of being removed, and then sprinkle the sand with water to cool and save it as much as possible. The consequence is that the part of the casting exposed to the air cools more rapidly than the part covered or partly covered by the sand, which creates a tension of the skin or outside of the casting. The same effect is produced, and to a greater extent, if water is sprinkled on one part of the casting and not on the other, or even on one part more than on another.

It has already been stated that brasses contract a little, sideways, in the process of boring, and that work of cast metal alters its form from the skin of the metal being removed; this alteration of form, in both cases, arises in the case of a piston ring from the release of the tension.

It sometimes occurs that a piece of work that is finished true in all its parts may unexpectedly require a cut to be taken off an unfinished part (to allow clearance or for other cause), and that the removal of the rough skin throws the work out of true in its various parts, as, for instance: a saddle of a lathe being scraped to fit the lathe bed, and its slides finely scraped to a surface plate; or the rest itself being fitted and adjusted to the cross slide of the saddle. If, when the nut and screw of the cross slide are placed in position, the nut is discovered to bind against the groove (of the saddle) along which it moves (the nut being too thin to permit of any more being taken off it), there is no alternative but to plane the groove in

TURNING PISTON RINGS. 149

the saddle deeper, which operation will cause the saddle to warp, destroying its fit upon the lathe bed, and the trueness of the V's of the cross slide, and that to such an extent as to sometimes require them to be refitted.

The evil effects of this tension may be reduced to a minimum by letting the casting cool in the mould, or if they are taken from the mould while still red hot, by placing them in a heap in some convenient part of the foundry, and covering them with sand kept in that place

Fig. 146.

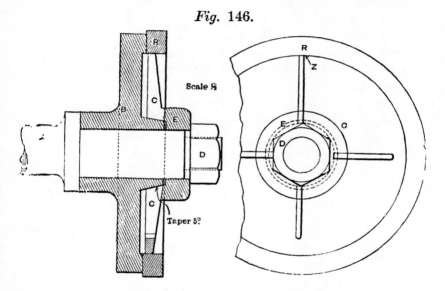

for the purpose; and by roughing out all the parts of the work which are to be cut at one chucking before finishing any one part.

Piston rings are turned larger than the bore of the cylinder which they are intended to fit, and, as before stated, sprung into the cylinder. The amount to which they are turned larger depends upon the form of split intended to be given to the ring; if it be a straight one, cut at an angle to the face of the ring, which is the form commonly employed, the diameter of the ring may be made in the proportion of one-quarter inch per foot larger than the bore of the cylinder, sufficient being cut out of the ring, on one side of the split, to permit the ring to spring into the diameter of the cylinder, when the ring may be placed in the cylinder and filed to fit, taking care to keep the ring true in the cylinder while revolving it to mark it.

Fig. 146 represents an expanding chuck for holding piston rings or similar work. It consists of an arbor or mandril A, upon which is the body B of the chuck, whose hub is coned. C is a disk bored to fit the coned hub of B, and having four splits, one of which, Z, extends to its circumference. E is a ring to receive the pressure of nut D, and it is obvious that if the latter be screwed up upon A, then disk C will be forced up the cone, and its diameter will enlarge and grip the bore of the ring or work R. The range of an expansion chuck of this kind is obviously small, hence it is suitable mainly for special work.

CHAPTER VII.

HAND-TURNING.

TURNING work in the lathe with a tool held or guided by hand, or, as it is commonly termed, hand-turning, is at once one of the most delicate and instructive branches of the machinist's art, imparting a knowledge of the nature and quantity of the resistance of metals to being cut, of the qualifications of various forms of cutting tools, and of the changes made in those qualifications consequent upon the relative position or angle of the cutting edge of the tool to the work; and this knowledge is to be obtained in no other way than by the practice of hand-turning.

It is the work of an instant only to vary the relative height and angle of a hand tool to the work, converting it from a roughing to a finishing tool or even to a scraper, which operations are difficult and sometimes impracticable, if not impossible, of accomplishment with a tool held in a slide rest.

The experience gained from the use of slide rest tools is imparted mainly through the medium of the eyesight, whereas in the case of a hand tool the sense of feeling becomes an active agent in imparting, at one and the same time, a knowledge of the nature of the work and the tool; so much so, indeed, that an excess in any of the requisite qualifications of a hand tool may be readily perceived from the sense of feeling, irrespective of any assistance from the eye; and in this fact lies the chief value of the experience gained by learning to turn by hand.

For instance, there is no method known to practice whereby to ascertain how much power it requires to force a slide rest tool into its cut, or to prevent its ripping in; so that a wide variation, in the tendency of such a tool to perform its allotted duty easily and without an unnecessary expenditure of power, may exist without becoming manifest to any save the experienced workman; whereas the amount of power required to keep the cutting edge of a hand tool to its work, to hold it steadily, or to prevent it from ripping, is communicated instantly to the understanding through the medium of the sense of feeling. Nor is this all, for even the sense of smell becomes a valuable assistant to the hand-turner. Several metals, especially wrought-iron, steel and brass, emit (when cut at a high speed) a peculiar smell, which becomes stronger with the increase in the speed at which they are cut and the comparative dulness of the edge of the tool employed to cut them, more especially when the cutting edge of the tool is supplied with oil during the operation of cutting. The reason that this sense of smell becomes more appreciable during the operation of hand than during that of slide rest turning, is because the face of the operator is nearer to the work, and because hand-turning is performed at a higher rate of cutting speed.

If a tool for use in a slide rest is too keen for i's allotted duty, the only result under ordinary circumstances is, that it will jar or chatter (that is, tremble and cut numerous indentations in the work), or that it will loose its cutting edge unnecessarily soon. But a hand tool possessing this defect will in many instances rip into the work, because the power, required to prevent the strain, placed by the cut upon the tool, from forcing the tool deeper into its cut than is intended, is too great to be sustained by the hand; and the tool, getting beyond the manipulator's control, rips into the work, cutting a gap or groove in it, and perhaps forcing it from between the centres of the lathe. If, on the other hand, a tool is of such a form that it requires a pressure to keep it to its duty, the amount of such pressure, when the tool is held at any relative height and angle to the horizontal centre line of the work, and the variation in that amount, due to the slightest alteration of the shape of the tool, are readily appreciated by sensitiveness of the hand; when they would be scarcely, if at all, perceived were the same tool, under like conditions, used in a slide rest.

These considerations, together with the great advantage in the relative rapidity with which the form and applied position of a hand tool may be varied, render hand-turning far more instructive to a beginner than any other branch of the machinist's art.

It is a common practice to centre one end of the work only, and to fasten the other end in a chuck, thus making the chuck serve as a driver, and obviating the necessity of centre-punching more than one end of the work. This method will, it is true, save a little time, but is objectionable for the following reasons: Chucks will run quite true while they are new, and indeed for some little time, but they do in time get out of true; and as a result, if the work requires to be reversed in the lathe so as to be turned from end to end, the part of the work turned during the second chucking will be eccentric to that part turned during the first chucking. If one end only of the work requires to be turned, and needs be true only of itself, and irrespective of the part held in the chuck, the latter may be employed; this subject will, however, be treated hereafter.

Our first operation, that is, truing the end of the work, is performed with a side tool, of which there are two kinds, both being made of three-cornered (or three-square, as it is generally termed) steel, the only point of difference being in the manner of grinding them. A worn-out saw file is an excellent thing to make a side tool of, because the teeth grip the rest and prevent the tool from slipping. It is not necessary to soften the file at all, but (for either kind) merely to grind it so as to make one edge a cutting one, and not make the point too thin, by grinding the end off a trifle.

If the cutting edges are smoothed by the application of an oilstone, they will give a very clean and smooth polish to the work. The rest should be set at such a height that the cutting edge of the tool is slightly above the horizontal centre of the work; and the tool should be so held that its side face stands nearly parallel with the end face of the work, the cutting edge being held slightly inclined towards the work, which will give to the tool edge the necessary clearance. Any excess of this inclination renders the tool liable to turn out of true, and destroys its cutting edge very rapidly.

ROUGHING OUT.

Our work, being countersunk, is now ready to be turned down to nearly the required size all over, before any one part is made to the finished size.

From what has been said in another place, the importance (in work which requires to be kept very true) of roughing the work out all over before any one part is finished will be obvious, since the breaking of the skin in any one part releases the tension on that part, whatever be the temperature it is under when in operation. It is not practicable, on lathe work, to at all times rough the work out all over before finishing any part; but in our present operation, of turning down a plain piece of iron held between the lathe centres, we are enabled to pursue that course, and we will therefore commence the roughing-out process with a graver.

THE GRAVER

is formed by grinding the end of a piece of square steel at an angle across the end, giving it a diamond-shaped appearance.

HAND-TURNING.

The graver is the most useful of all hand tools used upon metals. It can be applied to either rough out or finish steel, wrought-iron, cast-iron, brass, copper or other metal, and will turn work to almost any desired shape. Held with a heel pressed firmly against the hand rest (the point being used to cut, as shown in Fig. 147, A being the work, B the graver, and C the lathe rest), it turns very true, and cuts easily and freely. This, therefore, is the position in which the graver is held to rough out the work.

Fig. 147.

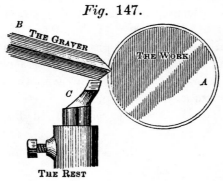

The heel of the graver, which rests upon the hand rest, should be pressed firmly to the rest, so as to serve as a fulcrum and at the same time as a pivotal point upon which it may turn to follow up the cut as it proceeds. The cutting point of the graver is held at first as much as convenient toward the dead centre, the handle in which the graver is fixed being held lightly by both hands, and slightly revolved from the right towards the left, at the same time that the handle is moved bodily from the left towards the right. By this combination of the two movements, if properly performed, the point of the graver will move in a line parallel to the centres of the lathe, because, while the twisting of the graver handle causes the graver point to move away from the centre of the diameter of the work, the moving of the handle bodily from left to right causes the point of the graver to approach the centre of that diameter; hence the one movement counteracts the other, producing a parallel movement, and at the same time enables the graver point to follow up the cut, using the heel as a pivotal fulcrum, and hence obviating the necessity of an inconveniently frequent moving of the heel of the tool along the rest. The most desirable range of these two movements will be very readily observed by the operator, because an excess in either of them destroys the efficacy of the heel of the graver as a fulcrum, and gives it less power to cut, and the operator has less control of the tool.

The handle in which the graver is held should be sufficiently long to enable the operator to grasp it with both hands and thus to hold it steadily, even though the work may run very much out of true.

To cut smoothly, as is required in finishing work,

Fig. 148.

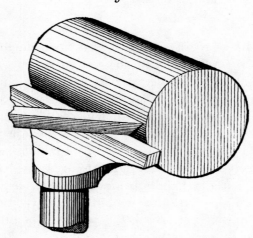

the graver is held as shown in Fig. 148, moving it from place to place along the work, and testing it for parallelism with the calipers. For finishing curves, however, the end face of the graver should be ground, curved from the heel to the point, but of less curvature than the work. Even parallel work should be finished by being filed with a smooth file while the lathe is running at a high speed. As little as possible should, however, be left for the file to do, because it cuts the softer veins of the metal more readily than the rest, and therefore makes the work out of true.

For use on brass and other soft metals, the two top flat sides of the graver should be ground away so as to have a negative top rake. The strain on the tool, when cutting soft metals, is comparatively slight, so that the graver is rarely applied to such metals in the position shown in Fig. 147.

THE HEEL TOOL.

In those exceptional cases in which, for want of a lathe having a slide rest, it becomes necessary to perform comparatively heavy work in a hand lathe, the heel tool should be employed. This tool was formerly held in great repute, but has become less useful by reason of the advent and universal application of the slide rest. It is an excellent one for roughing work out, and will take a very heavy cut for a hand tool, because of the great leverage it possesses, by reason of its shape and handle, over the work. A heel tool is shown in Fig. 149, A being the tool, which is a piece of square bar steel forged at the end to form the cutting edge. The body of the square part is held (in a groove formed in the wooden handle B) by an iron strap C, which is tightened by screwing up the under handle D, which contains a nut into which the spindle of the strap C is screwed as the handle D is revolved. The heel F of the tool is tapered, so that it will firmly grip the face of the lathe rest, the cutting edge E being rounded as shown in Fig. 149. The tool is held by grasping the handle B at about the point G, with the left hand, and by holding the under handle D in the right hand, the extreme end H of the handle being placed firmly against the right shoulder of the operator. The heel F of the tool must be placed directly under the part of the work it is intended to turn, the cutting edge E of the tool being kept up to the cut by using the handle D as a lever, and the heel F of the tool as a fulcrum. Not much lateral movement must, however, be allowed to the cutting edge of the tool to make it follow the cut, as it will get completely beyond the manipulator's control and rip into the work. Until some knowledge of the use of this tool has been acquired, it is better not to forge the top of

Fig. 149.

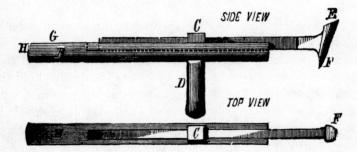

the cutting edge E too high from the body of the tool; since the lower it is the easier the tool is to handle.

The heel tool should, like the graver, be hardened right out; but in dipping it, allow the heel F to be a little the softer by plunging the end E into the water about half way to F; and then, after holding it in that position for about four seconds, immerse the heel F also. After again holding the tool still for about six seconds, withdraw it from the water and hold it until the water has dried off the point E; dip the tool again, and quickly withdraw it, repeating this latter part of the operation until the tool is quite cold. The object of the transient dippings is to prevent the junction of the hard and soft metal from being a narrow strip of metal, in which case the tool is very liable to break at that junction. The tool should be so placed in the handle that there is only sufficient room between the cutting edge and the end of the handle to well clear the lathe rest, and should be so held that the handle stands with the end H raised slightly above a horizontal position, the necessary rake being given by the angle of the top face at E. It is only applicable to wrought-iron and steel; but for use on those metals, especially the latter, it is a superior and valuable hand tool.

For cutting out a round corner, a round-nosed tool of the same description as the V tool given for starting threads by hand, but having the cutting edge ground round instead of a V shape, is the most effective; it will either rough out or finish, and may be used with or without water, but it is always preferable to use water for finishing wrought-iron and steel. This is a sample of a large class, applicable to steel and wrought-iron, the metal behind the cutting edge being ground away so as to give to the latter the keenness or rake necessary to enable it to cut freely, and the metal behind the heel being ground away to enable it to grip the rest firmly.

HAND-TURNING—BRASS WORK.

For roughing out brass work, the best and most universally applicable tool is that shown in Fig. 150, which is to

Fig. 150.

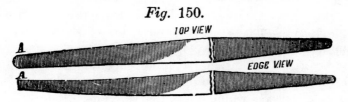

brass work what the graver is to wrought-iron or steel. The cutting point A is round-nosed. The hand rest should be set a little above the horizontal centre of the work, and need not be close up to the work, because comparatively little power is required to cut brass and other soft metals, and therefore complete control can be had over the tool, even though its point of contact with the rest be some little distance from its cutting point, which allows a greater range of movement of the tool from a fixed point. The best method of holding and guiding is to place the forefinger of the left hand under the jaw of the hand rest, and to press the tool firmly to the face of the rest by the thumb, regulating the height so that the cutting is performed at or a little below the horizontal centre of the work. The tool point may thus be guided with comparative ease to turn parallel, taper, or round or hollow curves, or any other desirable shape, except it be a square corner. Nor will it require much moving upon the face of the lathe rest, because its point of contact, being somewhat removed from the rest, gives to the tool point a comparatively wide range of movement. The exact requisite distance for the rest to be from the work must, in each case, be determined by the depth of the cut and the degree of hardness of the metal; but as a general rule, it should be as distant as is compatible with a thorough control of the tool. The cutting end of this tool should be tempered to a light straw color.

SCRAPERS.

To finish brass work, various shaped tools, termed scrapers, are employed. The term scraper, however, applies as much to the manner in which the tool is applied to the work as to its shape, since the same tool may, without alteration, be employed either as a scraping or a cutting tool, according to the angle of the top face (that is, the face which meets the shavings or cuttings) to a line drawn from the point of contact of the tool with the work to the centre line of the work, and altogether irrespective of the angles of the two faces of the tool whose junction forms the cutting edge. To give, then, the degree of angle necessary to a cutting tool, irrespective of the position in which it is held, is altogether valueless, as will be readily perceived.

Scrapers will cut more freely if applied to the work with the edges as left by the grindstone; but if they are smoothed, after grinding, by the application of an oilstone, they will give to the work a much smoother and higher degree of finish. They should be hardened right out for use on cast-iron, and tempered to a straw color for brass work. If the scraper jars or chatters, as it will sometimes, by reason of its having an excess of angle or bottom rake, or from the cutting end being ground too thin, a piece of leather, placed between the tool and the face of the rest, will obviate the difficulty.

Round or hollow curves may be finished truly and smoothly by simply scraping; but parts that are parallel or straight upon their outer surfaces should, subsequent to the scraping, be lightly filed with a smooth file, the lathe running at a very high speed to prevent the file from cutting the work out of true. The file should, however, be kept clean of the cuttings by either using a file card or cleaner, or by brushing the hand back and forth on the file, and then striking the latter lightly upon a block of wood or a piece of lead, the latter operation being much the more rapid, and sufficiently effective for all save the very finest of work. If the filings are not cleaned from the file, they are apt to get locked in the file teeth and to cut scratches in the work. To prevent this the file may be rubbed with chalk after every eight or ten strokes, and then cleaned as described. After filing the work, it may be polished with emery paper or emery cloth. The finer the paper and the more worn it is, the better and finer will be the finish it will give to the work; for all metals polish best by being rubbed at a high speed with a thin film composed of fine particles of their own nature, as ivory is best polished by ivory powder, and wood by shavings cut from itself. To facilitate obtaining the film of metal upon the emery paper, the latter may be oiled to a very slight extent, by rubbing a greasy rag over it, which will cause the particles it at first cuts to adhere to its surface. Crocus cloth is the best for highly finishing purposes, because it will wear longer without becoming torn. It should be pressed hard against the work, and reversed in all directions upon it, so as to wear all parts of its surface equally, and to distribute the metal film all over; and the work should be revolved at as high a speed as possible, while the crocus cloth, during the first part of the polishing, is kept in rapid motion upon the work backward and forward, so that the marks made upon the work by the emery cloth will cross and recross each other. When fine finishing is to be performed, the crocus cloth should be pressed very lightly against the work and moved laterally very slowly.

Round or hollow corners, or side faces of flanges, of either wrought or cast-iron or brass, may be polished with grain emery and oil, applied to the work on the end of a piece of soft wood, the operation being as follows: The end of the wood to which the oil and emery is to be applied should be slightly disintegrated by being bruised with a hammer; this will permit the oil and emery to enter into and be detained in the wood instead of passing away at the sides, as it otherwise would do, thus saving a large proportionate amount of material. The wood, being bruised, will also conform itself much more readily to the shape of curves, grooves, or corners. The hand rest is then placed a short distance from the work, and the piece of wood rests upon it, using it as a fulcrum. The end of the wood should bear upon the work below the horizontal level of the centre of the latter, so that depressing the end of the wood held in the hand employs it as a lever, placing considerable pressure against the work; and the distance of the rest from the work allows the end of the piece of wood to have a reasonable range of lateral movement, without being moved upon the face of the lathe rest. The method of using the wood is the same as that employed in using emery cloth, except that it must, during the earlier stage of its application, be kept in very continuous lateral movement, or the grain emery will lodge in any small hollow specks which may exist in the metal, and hence cut small grooves in the work. Another exception is that the finishing must be performed with only such emery as may be embedded in the wood, and without the application of any oil; especially are these directions necessary for cast-iron or brass work. The work may then be wiped dry, and an extra polish imparted to it by the application of fine or worn and glazed emery cloth, moved slowly over its surfaces.

CHAPTER VIII.

DRILLING IN THE LATHE.

WE have next to consider drilling tools as they are employed in the lathe. For boring very small holes, as in centre-drilling, it is usual and advisable to revolve the drill and use the dead centre and its gear as a feed motion. For small lathes, a small chuck or face plate is made, it having a conical stem so as to fit into the hole into which the dead centre fits.

It is obvious that, as a lathe possesses no facilities for chucking work upon the tail stock, work which requires chucking, or is too heavy to be held conveniently in the hand, can only be drilled in the lathe by being chucked and revolved, the drill remaining stationary, and fitted into the socket in the tail stock spindle, or else suspended by being held by the work at the cutting end, and by the dead centre at the other end, and prevented from revolving by the aid of a drilling rest or a wrench. If the work revolves, it must of course be set to run true; and since the setting involves more work than would be required to hold it upon a drilling machine table, it follows that the lathe is only resorted to for drilling purposes in cases in which it is imperative to use it. These instances may be classified as follows:

1. Those in which very straight and true holes are required, and in which the point of ingress and egress may be centre-punched, in which cases (the back centre of the lathe being placed in the centre punch mark, and the point of the drill in the other) the drilling is sure to be true.

DRILLING IN THE LATHE.

2. Those in which the work being very long, can be got into the lathe in consequence of the movable tail stock, when it could not be got into the drilling machine.

3. Those in which, there being turning to be done besides the boring or drilling, the whole may be performed in the lathe.

4. Those in which the holes require to be very true, the work being chucked in the lathe.

The class first mentioned refers to small and light work only, and requires no comment, save that the work should be slowly revolved on the lathe centre while the drilling is progressing, so that the work will not drill out of true in consequence of its weight. The second was referred to under the heading of the cone plate, or cone chuck, as it is sometimes termed; and the third (which usually comprises the fourth) we will proceed to discuss.

The spindle in the tail stocks of lathes are usually prevented from revolving by having a narrow groove along them, into which a small lug, stationary with, and projecting through, the bearing of the spindle, fits. If, therefore, a heavy strain, tending to twist the socket (as would be the case if a drill of a comparatively large size were held by it), is placed upon it, the groove, from its comparatively small wearing surface, soon gets worn as well as the lug, and the edge of the groove bulges, causing the socket to bind in its guide. Tail stock spindles are not, in fact, usually designed to perform such heavy duty; hence it is an error to assign it to them, unless, as is the case in some special lathes, the tail stock spindles, and hence their bearings, are made square to suit the spindles to carry drills for heavy duty. For ordinary drilling in the lathe the twist drill is employed, but since it is used in the drilling machine also, it will be considered in connection with other drilling tools, and we may therefore pass to such tools as are used more exclusively on lathe work, whether for drilling holes out of the solid metal, or for enlarging holes that have been cast or forged in the work and which are used upon work that is chucked upon the face plate or in other chucking devices.

HALF ROUND BITS.

For drilling or boring holes very true and parallel in the lathe, the half round bit shown in Fig. 151 is unsurpassed.

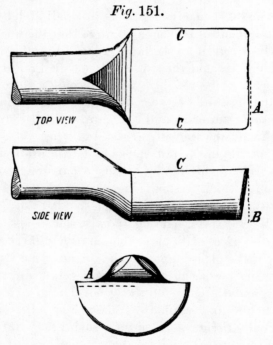

Fig. 151.

The cutting edge A is made by backing off the end, as denoted by the space between the lower end of the tool and the dotted line B, and performing its duty along the radius, as denoted by the dotted line in the end and top views.

It is only necessary to start the half round bit true, to insure its boring a hole of any depth, true, parallel, and very smooth. To start it, the face of the work should, if

DRILLING IN THE LATHE.

the centre upon which the tool has been turned, which line will form a guide for filing the top face down to make the tool of the required thickness of one-half of its diameter. The edge A should be perfectly square with the side or diametrical edges C C. The circumference of the turned part should have the turning marks effaced with a very smooth file, by draw-filing the work lengthwise, care being taken to remove an even quantity all over. The rake of the tool, as denoted at the dotted line B, should not be greater in proportion than is there shown.

This tool should be tempered to a straw color and employed at a cutting speed of about fifteen feet per minute, and fed at a coarse feed by hand. For use on parallel holes, no part should be ground save the end face; whereas, in the case of taper ones, the top face may be ground, taking a little off as will answer the purpose. It should be borne in mind that, as the steel expands (and therefore becomes larger in diameter) by the process of hardening, the necessary allowance, which is about the one-fiftieth of an inch per inch of diameter, should be made when turning it in the lathe. Tools of this description, which have a turned part to guide them, or those which depend upon the trueness of their outline or cutting edges to make them perform their duty, and which are apt, in the process of hardening, to get out of true (for all steel alters more or less during the operation of hardening), may be made true after the hardening or tempering by a process to be described in our future remarks on reamers, since it applies more directly to those tools than to half round bits.

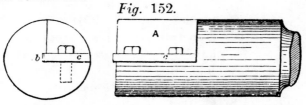

Fig. 152.

Fig. 152 represents a bit in which a segment A is

cut out to admit a cutter C, which may be adjusted to size by slips of paper put in at C.

Figs. 153 and 154 represent a side and an end view of a

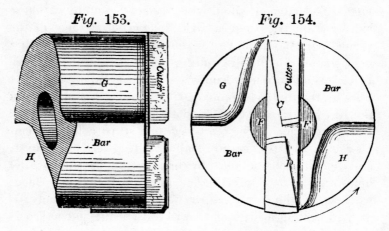

Fig. 153.   Fig. 154.

cutter and bar (and Fig. 155 a side view of the cutter removed from the bar), especially designed for piercing holes out of the solid, and of great depth. The cutting edges C and D form a radial line, and the latter does not extend to the centre. As a result, there is formed a slightly projecting edge to the work, which acts as a guide to keep the cutter true. The end A of the cutter fits into the bore of the bar, and the latter is provided with longitudinal grooves G H, so that water forced through the bore of the bar will wash the cuttings out through the grooves G H.

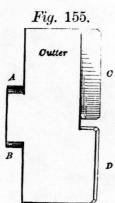

Fig. 155.

To enlarge holes and true them out, the flat drill (Fig. 156) is employed. It is an ordinary drill made out of flat steel, having pieces of hard wood fastened to the cutting end, A being the steel, and B B pieces of wood, held on by screws. When the drill has entered the hole far enough to make it of the diameter of the drill, the

DRILLING IN THE LATHE.

pieces of wood enter and fit the hole, steadying the drill and tending to keep it true. It is necessary, however, to true out the hole at the outer end before inserting the drill; for if the drill enters out of true, it will get worse as the work proceeds. The drill is fed to its duty by the back lathe centre, placed in the centre upon which the drill has been turned up.

The pieces of wood should be affixed before the drill is turned up, and so trued up with the drill, which should then be lightly draw-filed on the sides; and the cutting end having the necessary rake filed upon it, should be

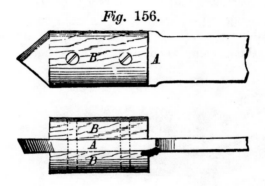

Fig. 156.

tempered to a straw color, the pieces of wood being, of course, temporarily removed. For use on conical holes, the sides must be made of the requisite cone and the cutting speed in that case reduced (in consequence of the broad cutting surface) to about 10 feet per minute. (This speed will also serve in boring conical holes with a half round bit.) Such a drill is an excellent tool for ordinary work, such as pulleys, etc., because it will perform its duty very rapidly and maintain its standard size; and it requires but little skill in handling. It is more applicable, however, to cast-iron than to any other metal. After the outer end of the hole has been turned true and of the required size, to receive the drill, and when the latter is

15

inserted for operation, it is an excellent plan to fasten a piece of metal, such as a lathe tool, into the tool post, and adjust the rest so that the end of the tool has light contact with the drill, so as to steady it. The lathe should be started, and the tool end wound in by the screw of the rest, until, the drill being true, the tool end just touches it, and having its end bevelled so as to have contact with the drill as close to the entrance of the hole as possible, in which position it is most effective. In all cases, when a drill is used in the lathe and remains stationary while the work revolves, this steadying implement should be employed, since it operates greatly to correct any tendency of the drill to spring out of true.

To hold flat drills, or those having square ends, and prevent them from revolving, a drill holder may be

Fig. 157.

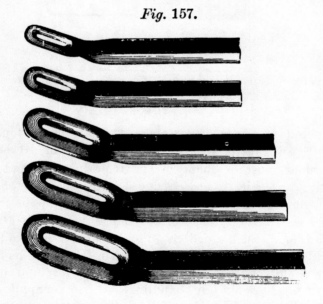

employed, either at the front end of the drill immediately behind the wood, or at the other end near the dead centre, the shape of the holder being as shown in Fig. 157, which shows five sizes. The angle of the eye to the body of the bar being so that the slide rest will stand off and not be close up to the chuck plate or the end of the work. It is well to keep the eye of the drill holder close to the entrance of the hole being drilled.

Fig. 158.

REAMERS.

The reamer consists of a hardened piece of steel, fluted as shown in Fig. 158, so as to produce cutting edges at the tops of the flutes. It is revolved and forced endways into the work.

The reamer owes its present state of perfection to the emery-wheel, which grinds it true after the hardening process, and the main considerations in determining its form are as follows:

1. The number of its cutting edges.
2. The spacing of the teeth.
3. The angles of the faces forming the cutting edges.
4. Its maintenance to standard diameter.

As to the first, it is obvious that the greater the number of cutting edges the more lines of contact there are to steady it on the walls of the hole; but in any case there should be more than three teeth, for if three teeth are used, and one of them is either relieved of its cut, or takes an excess of cut by reason of imperfections in the roundness of the hole, the other two are similarly affected and the hole is thus made out of round.

As to the spacing of the teeth, it is determined to a great

extent by the size of the reamer and the facilities that size affords for grinding the reamer.

The method employed to grind a reamer is shown in Fig. 159, in which is represented a rapidly revolving emery-wheel, a reamer, and also a gauge against which the front face of each tooth is held while its top or circumferential face is being sharpened. The reamer is held true to its axis, and is pushed endways beneath the revolving emery-wheel. In order that the wheel may leave the right-hand or cutting edge the highest (as it must be to enable it to cut), the centre of the emery-wheel

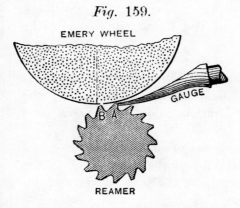

Fig. 159.

must be on the left hand of that of the reamer, and the spacing of the teeth must be such that the periphery of the emery-wheel will escape tooth B, for otherwise it would grind away its cutting edge. It is obvious, however, that the less the diameter of the emery-wheel, the closer the teeth may be spaced; but there is an objection to this, inasmuch as that the top of the tooth is naturally ground to the curvature of the wheel, as is shown in Fig. 160, in which two different-sized emery-wheels are represented, operating on the same diameter of reamer. The cutting edge of A has the most clearance, and is therefore the weakest and least durable; hence it is desirable to employ as large a wheel as the spacing of the teeth will allow, there being at least four teeth, and preferably six, on small reamers, and their number increasing with the diameter of the reamer.

Fig. 160.

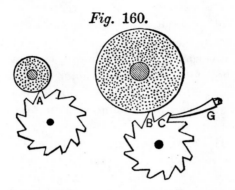

Concerning the angles of the faces forming the cutting edges, it is found that the front faces, as A and B in Fig.

Fig. 161.

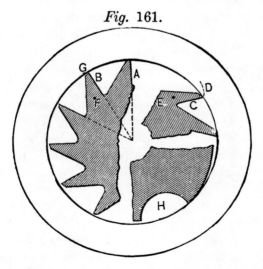

161, should be a radial line, for, if given rake as at C, the tooth will spring off the centre at point E in the direction of D, and cause the reamer to cut a hole of larger diameter

15*

than itself, an action that is found to occur to some extent even where the front face is a radial line. As this spring augments with any increase of cut-pressure, it is obvious that if a number of holes are to be reamed to the same diameter, it is essential that the reamer take the same depth of cut in each, so that the tooth-spring may be equal for each. The clearance at the top of the teeth is obviously governed by the position of the reamer with relation to the wheel, and the diameter of the wheel, being less in proportion as the reamer is placed farther beneath the wheel, and the wheel diameter is increased. In some forms of reamer the teeth are formed by circular flutes, such as at H in Fig. 161, and but three flutes are used. This leaves the teeth so strong and broad at the base that the teeth are not so liable to spring; but, on the other hand, the clearance is much more difficult to produce and to grind in the resharpening; hence such reamers have not found favor in the United States.

As to the maintenance of the reamer to standard diameter, it is a matter of great importance, for the following reasons: The great advantage of the standard reamer is to enable holes to be made and pieces to be turned to fit in them without requiring any particular piece to be fitted to some particular hole, and in order to accomplish this it is necessary that all the holes and all the pieces be exactly alike in diameter. But the cutting edges of the reamer begin to wear—and the reamer diameter, therefore, to reduce—from the very first hole it reams, and it is only a question of time when the holes will become too small for the turned pieces to enter or fit properly. In all pieces that are made a sliding or a working fit, as it is termed when one piece moves upon the other, there must be allowed a certain latitude of wear before the one piece must be renewed.

One course is to make the reamer, when new, enough larger than the proper size, to bore the holes as much

ADJUSTABLE REAMERS. 175

larger as this limit of wear, and to restore it to size when it has worn down so that the holes fit too tightly to the pieces that fit them. But this plan has the great disadvantage that the pieces generally require to have other cutting operations performed on them after the reaming, and to hold them for these operations it is necessary to insert in them tightly fitting plugs, or arbors, as they are termed. If, therefore, the holes are not of equal diameter, the arbor must be fitted to the holes, whereas the arbor should be to standard diameter to save the necessity of fitting, which would be almost as costly as fitting each turned piece to its own hole. It follows, therefore, that the holes and arbors should both be made to a certain standard, and the only way to do this is to so construct the reamer that it may be readily adjusted to size by moving its teeth, a reamer so constructed being shown in Fig. 162. The stock is, it will be seen, provided with dovetail grooves that are deepest towards the point, so, that by moving the teeth towards the shank, their diameter is increased.

Fig. 163 represents an adjustable reamer for very small work. It is pierced with a tapped hole and countersunk, and is split through at the end. A small plug P is inserted, and a screw S, and it is obvious that by screwing in the plug, and then the screw, the diameter of the reamer is enlarged. The pressure between the plug P and screw S serves to lock the latter in its adjusted position. It is obvious that the split weakens the reamer, hence it is only suitable for finishing to size.

Fig. 162.

SHELL REAMERS.

Shell reamers, such as shown in Fig. 164, are excellent tools for sizing purposes; that is, for taking a very light cut intended merely to smooth out the hole, and insure correctness in its bore or size. The notch fits a pin in the mandril and prevents it from slipping upon the mandril as it is otherwise very apt to do.

In the adjustable reamer, shown in Fig. 165, A represents the stock and D the cutter, C being a regulating washer, and D and E the tightening nut and washer. Each of the cutters B fits into a dovetail and taper

Fig. 163.

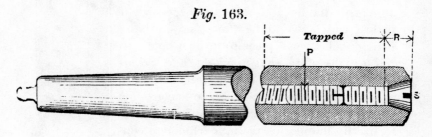

groove in the stock, the shallow end of the groove being at the cutting end; so that if the regulating washer C is reduced in width, the cutters will slide forward and enlarge in diameter. The washer C is thus a means of adjusting the diameter of the cutters; and when the same is once adjusted, the nut D will lock it always to that precise diameter. If, therefore, several sets of cutters of different heights are fitted to one stock, and turned up while in the stock to the requisite diameter with the washer C in its place, we have a set of standard cutters which may always be placed in position

Fig. 164.

and locked up by the nut D, without measurement, since their sizes cannot vary. By providing another washer, very slightly thicker than the standard, the reamer will, in the case of each set of cutters, bore a hole to a driving fit, while a washer a trifle thinner will cause the cutters to bore a hole of an easily working fit. Thus the sizes of the cutters are regulated by the washer C, and not by measurement by the workman; they are therefore at all

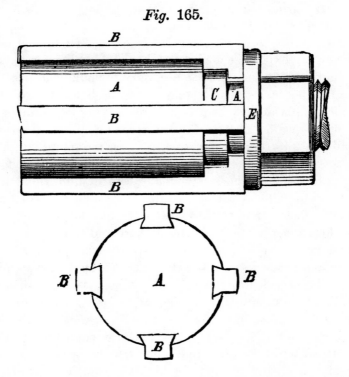

Fig. 165.

times positive and equal. The cutters are backed off on the ends only, their tops being merely lightly draw-filed after being turned up, or they may be left one thirty-second of an inch too large, and ground off after hardening, by the grinding process already described. The cutters should be forged of the best cast-steel and tempered to a straw color.

CHAPTER IX.

BORING BARS.

The boring bar is one of the most important tools to be found in a machine shop, because the work it has to perform requires to be very accurately done; and since it is a somewhat expensive tool to make, and occupies a large amount of shop room, it is necessary to make one size of boring bar answer for as many sizes of hole as possible, which end can only be attained by making it thoroughly stiff and rigid. To this end a large amount of bearing and close fitting, using cast-iron as the material, are necessary, because cast-iron does not spring or deflect so easily as wrought-iron; but the centres into which the lathe centres fit are, if of cast-iron, very liable to cut and shift their position, thus throwing the bar out of true. It is, therefore, always preferable to bore and tap the ends of such bars, and to screw in a wrought-iron plug, taking care to screw it in very tightly, so that it shall not at any time become loose. The centres should be well drilled and of a comparatively large size, so as to have surface enough to suffer little from wear, and to well sustain the weight of the bar. The end surface surrounding the centres should be turned off quite true to keep the latter from wearing away from the high side, as they would do were one side higher than the other.

The smaller sizes of boring bars are usually simple parallel mandrils, having slots running through them, into which slots or keyways the cutters are fitted, being fastened by means of wedges. The backs of the cutters are tapered to the same degree as is the wedge, so that the key will bear evenly along both the edge of the keyway and the cutter. It is obvious that, if the cutter is turned up in the bar, and is of the exact size of the hole to be bored, it will require to stand true in the bar, and will therefore be able to cut on both ends, in which case the work may be fed up to it twice as fast as though only one edge were performing duty. To facilitate setting the cutter quite true, a flat and slightly taper surface should be filed on the bar at each end of the keyway, and the cutter should have a recess filed in it to fit the diameter of the bar so filed, so that after passing the cutter through the slot, it may be pushed forward in the manner of a jib, and then locked by the wedge. Such cutters not being adjustable, their diametrical edges need not have any clearance or rake on them, but the cutting corners should be rounded off, and the rake put on the end face of the cutter and carried around the round corner, the advantage being that the diametrical edge of the cutter will bear lightly against the bore of the work, and prevent the bar from springing.

Boring bar cutters, required to be adjustable, must not be provided with a recess, but must be left plain, so that they may be made to extend out on one side of the bar to cut any requisite size of bore; it is far preferable, however, to employ the recess and have a sufficient number of cutters to suit any size of hole, since, as already stated (there being in that case two cutting edges performing duty), the work may be fed up twice as fast as in the former case, in which only one cutting edge operates. This description of bar for use on small holes or bores is simply a mandril, and may be provided with several slots or keyways in its length, to facilitate facing off the ends of work which requires it. Since the work is fed to the cutter, it is obvious that the bar must be at least twice the length of the work, because the work is all on one side of the cutter at the commencement, and all on the other side at the conclusion of the boring operation. The excessive length of bar, thus rendered necessary, is the principal objection to this form of boring bar, because of its liability to spring. There should always be a keyway, slot, or cutter way in the exact centre of the length of the bar, so as to enable it to bore a hole as long as possible in proportion to the length of the boring bar, and a keyway or cutter-way at each end of the bar, for use in facing off. If, however, a boring bar is to be used for a job which does not require to be faced off at the ends, the keyway should be placed in such a position in the length of the bar as will best accommodate the work, and should then be made tapering in diameter from the keyway to the ends, a short piece at one end of the bar being made parallel to receive the driving clamp. A lug, however, by which to drive the bar, is sometimes cast on one end. This form of bar is stronger in proportion to its weight, and therefore less liable to spring from the cut or to deflect than is a parallel bar. The deflection of a bar, the length of which is excessive in proportion to its diameter, is sufficient to cause it to bore a hole out of straight in the direction of the length of the bore, providing that the cutter is not recessed and does not cut on both sides—that is to say, when the cutter has the diametrical bearing against the diameter of the hole, they serve to steady the bar and prevent it from either springing away from the cut, or from deflecting in consequence of its own weight. The question of spring affects all boring bars; but in those which are used vertically, the deflection is of course obviated.

Here it may be mentioned that no machine using a boring bar should be allowed to stop while the finishing cut is being taken, for the following reasons: The friction, due to the severance of the metal being cut, causes it to heat to a slight degree, and to therefore expand to an appreciable extent; so that when the cutter makes its first revolution, it is operating upon metal at its normal temperature, but the heat created has expanded the bore of the work, and hence the cut taken by the second revolution of the cutter will be slightly less in diameter. This heating and expanding process continues as the cutting proceeds, so that if (after the cutter has made any number of revolutions) the bar is stopped and the cylinder or other work being bored becomes cool, when the cutter makes the next revolution it will be operating upon the bore unexpanded by the heat, and hence will cut deeper into the metal, until the metal, being reheated by the cut during the revolution, the boring proceeds upon expanded metal as before the stoppage; thus arresting the continuous progress of the cutter will have caused the cutting of a groove in the bore. Boring bars, for use in bores of a large diameter, are made with a head of increased diameter, the body of the bar being turned along its length and provided with a slot or key groove from end to end, the sliding head is bored to fit the bar, and is provided with a keyway. Thus the head may be keyed to the bar at any part of the length of the latter. Several cutters may be provided to the head, so that the work may be fed up rapidly; in such case, however, great exactitude is required in setting them, because there is no practical method of making them with a recess to insure their even projection from the bar, since the cutters are narrow, and generally cut across the whole diametrical face, so that each grinding affects their distance from the bar, and hence the size they bore.

A rude form of head may be made by simply cutting a slot or slots across it, and fastening the tool or tools therein, by means of wedges, and packing pieces, if necessary. The only advantage possessed by this kind of bar is that it will bore a round hole, even though the bar may run out of true, by reason of either or both of the centres being misplaced, or even though the bar itself may have become bent in its length. In addition, however, to its disadvantage as to excessive length, it possesses the further one that, unless a line drawn from the two centres upon which it revolves is parallel both perpendicularly and horizontally to the lathe bed, the hole bored will be oval and not round; or if the bar is not parallel horizontally with the shears, the hole will be widest perpendicularly, and *vice versâ*. To remedy these defects, we have the boring bar with the feeding head, which is similar to that described, save that the work remains stationary while the cutters are fed to the work by operating the head along the bar, which is accomplished as follows: either along the keyway or groove, or else through and along the centre of the boring bar, there is provided a feeding screw, passing through a nut which is attached to the sliding head. As the bar revolves upon its axis, the screw is, by means of suitable gearing, caused to revolve upon its own axis, as well as around the axis of the bar, thus winding the head along the length of the bar, and thus feeding it to the cut. If the screw runs along the centre of the bar, it is usually operated by gear wheels, the movement of the feed being continuous at all parts of the revolution; but if the screw is contained in a groove cut in the circumference of the bar, a common star feed may be attached to the end of the bar, in which case the feed of the whole revolution is given to the sliding head during that portion only of the revolution in which the outer arm of the star is moved by the projecting bolt or arm which operates it. From these directions, it will be readily perceived that a bar of the latter form, but having the screw in its centre, is the most preferable. Care must be taken, however, to keep these bars running quite true; for should either centre run out of true, the hole bored will be larger in diameter at that end; while on the other hand, should the bar become bent so as to run out of true in the middle of its length, the hole bored will be large in the middle if the work was chucked in the middle of the length of the bar; and otherwise it will be larger at one end.

A very important consideration with reference to boring bars is the position which the cutters should occupy towards the head or the body of the bar. We have already been over the same ground with reference to parting or grooving tools for lathe work, cutting tools for planing work, and cutters for cutting out holes of a large diameter in boiler plates; but there are so many principles involved in the shape and holding position of cutting tools, so many variations, and so many instances in which the reasons for the adoption or variation of a principle are not obvious, that it is of vital importance to specify, in the case of each tool, its precise shape and position of application, together with the reasons therefor, the field of application being so extensive that the memory can hardly be relied upon. A careful survey of all the tools thus far treated upon will disclose that, in each case wherein the cutting edge stands in advance (in the direction in which the tool is moving, or, if the work move, in the direction of the metal to be cut) of the fulcrum upon which the tool is held, the springing of the tool causes it to dig into the work, deepening the cut, and in most cases causing the tool point or cutting edge to break; while in every instance this defect has been cured (upon tools liable to spring) by so bending or placing the tool that the fulcrum upon which it was held stood in advance of the cutting edge; and these rules are so universal that it may be said that pushing a tool renders it liable to spring into the work, and pulling it or dragging it enables it to take a greater cut and to spring away from excessive duty; and thus the latter prevents breakage and excessive spring, because, when the spring deepens the cut, it increases proportionally the causes of the spring, and creates a contention between the strength of the tool and the driving power of the machine, resulting in a victory for the one or the other, unless the work itself should give way, either by springing away from the tool and bending, or forcing it from the lathe centres or from the clamps which hold it.

For instance, in Fig. 166 is shown A, a boring bar; B B is the sliding head; C C is the bore of the cylinder, and 1, 2, and 3 are tools in the positions shown. D D D are projections in the bore of the cylinder, causing an excessive amount of duty to be placed upon the cutters, as sometimes occurs when a cut of medium depth has been started. Such a cut increases on one side of the bore of the work until, becoming excessive, it causes the bar to tremble and

Fig. 166.

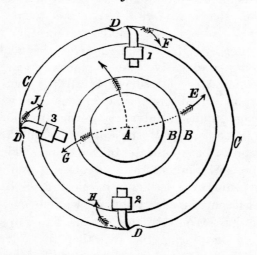

the cutters to chatter. In such a case, tool and position No. 1 would not be relieved of any duty, though it spring to a considerable degree; because the bar would spring in the direction denoted by the dotted line and arrow E, while the spring of the tool itself would be in the direction of the dotted line F. The tendency of the spring of the bar is to force the tool deeper into the cut instead of relieving it; while the tendency of the spring of the tool will scarcely affect the depth of the cut. Tool and position No. 2 would cause the bar to spring in the direction of the dotted line and arrow G, and the tool itself to spring in the direction of H, the spring of the bar being in a direc-

BORING BARS.   185

tion to increase, and that of the tool to diminish, the cut. Tool and position No. 3 would, however, place the spring of the bar in a direction which would scarcely affect the depth of the cut, while the spring of the tool itself would be in a direction to give decided relief by springing away from its excessive duty. It must be borne in mind that even a stout bar of medium length will spring considerably from an ordinary roughing-out cut, though the latter

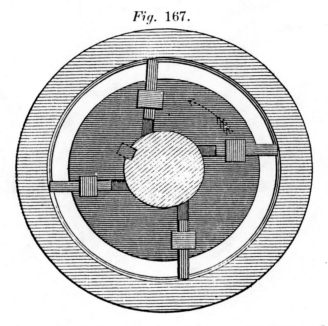

Fig. 167.

be of an equal depth all round the bore and from end to end of the work. Position No. 3, in Fig. 166, then is decidedly preferable for the roughing-out cuts. In the finishing cuts, which should be very light ones, neither the bar nor the tool are so much affected by springing; but even here position No. 3 maintains its superiority, because, the tool being pulled, it operates somewhat as a scraper (though it may be as keen in shape as the other tools) and hence it cuts more smoothly.

Fig. 167 represents a boring bar, the cutters standing to one side of the bar axis so as to carry out the principle explained with reference to figure 166.

Fig. 168 represents a boring bar having three cutters, and it will be seen that if one cutter, as A, leaves its cut, the pressure of the cut on the other two will spring the bar towards A, and the whole will not be round.

To obtain the very best and most rapid result, there should be but little space between the sliding head and

Fig. 168.

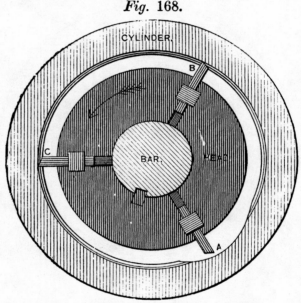

the bore of the work; the bar itself should be as stout as is practicable, leaving the sliding head of sufficient strength; and if the bar revolves in journals, these should be of large diameter and with ample facilities for taking up both the diametrical and end play of the boxes, since the one steadies the bar while it is performing boring duty, and the other while it is facing off end faces, as for cylinder cover joints. The feed of a boring bar, which is slight in comparison to its duty, will range at from twenty to thirty revolutions to an inch of travel; while that of a stout bar, held in large and closely-fitting journals, may be about sixteen revolutions per inch of tool travel for roughing-out cuts, and four revolutions per inch of travel for finishing cuts, which may be made to leave the work very smooth indeed.

The tools employed for the roughing cuts should not have a broad cutting surface, and should have a little front rake. For the finishing cuts, the same tool may be employed, the end being ground to have, for use on cast-iron, a broad, level cutting surface along the cutting edge, so that, while the front edge of the tool is cutting, the behind part will scrape and thus smooth the cut. These tools should be made of the best quality of steel, and hardened right out, that is to say, not tempered at all.

The lip or top rake must, in case the bar should tremble during the finishing cut, be ground off, leaving the face level; and if, from the bar being too slight for its duty, it should still either chatter or jar, it will pay best to reduce the revolutions per minute of the bar, keeping the feed as coarse as possible, which will give the best results in a given time. In cases where, from the excessive length and smallness of the bar, it is difficult to prevent it from springing, the cutters must be made with no lip, and but a small amount of cutting surface; and the corner A should be bevelled off as shown. Under these conditions, the tool is the least likely to chatter or to spring into the cut, especially if held in position No. 3, in Fig. 166, for a tool which would jar violently in position No. 1 would cut smoothly and well if held in position No. 3.

The shape of the cutting corner of a cutter depends entirely upon the position of its clearance or rake. If the edge forming the diameter has no clearance upon it, the cutting being performed by the end edges, the cutter may be left with a square, slightly rounded, or bevelled corner; but if the cutter have clearance on its outside or diametrical edge, as shown on the cutters in Fig. 166, the cutting corner should be bevelled or rounded off, otherwise it will jar in taking a roughing cut, and chatter in taking a moderate cut. The principle is, that bevelling off the front edge of the cutter tends greatly to counteract a disposition to either jarring or chattering, especially as applied to brass work.

The only other precaution which can be taken to prevent, in exceptional cases, the spring of a boring bar is to provide a bearing at each end of the work, as, for instance, by bolting to the end of the work four iron plates, the ends being hollowed to fit the bar, and being so adjusted as to barely touch it; so that, while the bar will not be sprung by the plates, yet, if it tends to spring out of true, it will be prevented from doing so by contact with the hollow ends of the plates, which latter should have a wide bearing and be kept well lubricated.

It sometimes happens that from play in the journals of the machine, or from other causes, a boring bar will jar or chatter at the commencement of a bore, and will gradually cease to do so as the cut proceeds and the cutter gets a broader bearing on the work. Especially is this liable to occur in using cutters having no clearance on the diametrical edge; because, so soon as such a cutter has entered the bore for a short distance, the diametrical edge (fitting closely to the bore) acts as a guide to steady the cutter. If, however, the cutter has such clearance, the only perceptible reason is that the chattering ceases as soon as the cutting edge of the tool or cutter has lost its fibrous edges. The natural remedy for this would appear to be to apply the oil-stone; this, however, will either have no effect or make matters worse. It is, indeed, a far better plan to take the tool (after grinding) and rub the cutting edge into a piece of soft wood, and to apply oil to the tool during its first two or three cutting revolutions. The application of oil will often remedy a slight existing chattering of a boring bar, but it is an expedient to be avoided, if possible, since the diameter or bore cut with oil will vary from that cut dry, the latter being a trifle the larger.

The considerations, therefore, which determine the shape of a cutter to be employed are as follows: Cutters for use on a certain and unvarying size of bore should have no clearance on the diametrical edges, the cutting being performed by the end edge only. Cutters intended to be adjusted to suit bores of varying diameter should have clearance on the end and on the diametrical edges. For use on brass work, the cutting corner should be rounded off, and there should be no lip given to the cutting edge. For wrought-iron the cutter should be lipped, and oil or soapy water should be supplied to it during the operation. A slight lip should be given to cutters for use on cast-iron, unless, from slightness in the bar or other causes, there is a tendency to jarring, in which case no lip or front rake should be given.

SMALL BORING BARS.

In boring work chucked and revolved in the lathe, such, for instance, as axle boxes for locomotives, the device shown in Fig. 169 is an excellent tool. A represents a

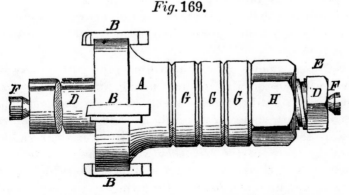

Fig. 169.

cutter head, which slides along, at a close working fit, upon the bar D D, and is provided with the cutters B B B, which are fastened into slots provided in the head A, by the keys shown. The bar D D has a thread cut upon part of its length, the remainder being plain, to fit the sliding head. One end is squared to receive a wrench, which, resting against the bed of the lathe, prevents the bar from revolving upon the lathe centres F F, by which the bar is held in the lathe. G G G are plain washers, provided to make up the distance between the thread and plain part of the bar, in cases where the sliding head A requires considerable lateral movement, there being more or fewer washers employed according to the distance along which the sliding head is required to move. The edges of these washers are chamfered off to prevent them from burring easily. To feed the cutters, the nut H is screwed up with a wrench.

The cutter head A is provided in its bore with two feathers, which slide in grooves provided in the bar D D, thus preventing the head from revolving upon the bar. It is obvious that this bar will, in consequence of its rigidity, take out a much heavier cut than would be possible with any boring tool, and furthermore that, there being four cutters, they can be fed up four times as fast as would be possible with a single tool or cutter. Care must, however, be exercised to so set the cutters that they will all project true radially, so that the depth of cut taken by each will be equal, or practically so ; otherwise the feeding cannot progress any faster than if one cutter only were employed.

For use on bores of a standard size, the cutters may be made with a projecting feather, fitting into a groove provided in the head to receive them. The cutters should be fitted to their places, and each marked to its place; so that, if the keyways should vary a little in their radius from the centre of the bar, they will nevertheless be true when in use, if always placed in the slot in which they were turned up when made. By fitting in several sets of cutters and turning them up to standard sizes, correctness in the size of bore may be at all times insured, and the feeding may be performed very fast indeed.

CHAPTER X.

SLOTTING MACHINE TOOLS.

Tools for use in slotting machines are divided into two classes: those used by themselves, for holes in which there is not sufficient room to admit a tool-post or bar; and short tools, held in a tool-post on the bar, and fastened by a set screw or screws thereon provided.

Referring to the first class, it is advantageous to let the cutting edge of the tool stand below the level of the bottom of the tool steel, so that in springing or deflecting from the pressure of the cut, which it is sure to do in some degree, the tool point will enter deeper into its cut, and will not, therefore, rub against it during the back or return stroke of the tool, as it is apt to do if the tool edge is level with the bottom of the tool steel.

If the tool edge is level with the middle of the tool steel, the spring or deflection, due to the strain or pressure of the cut, causes the cutting edge to spring away from the work, and, therefore, lessen the depth of the cut; hence during the back stroke the tool, being relieved of strain, rubs against the side of the cut, and the abrasion rapidly dulls the cutting edge. Similarly by giving the front face side rake, the tool will move slightly in the direction of the feed during the cutting stroke, and be correspondingly relieved during its return stroke. When the tool is slight and stands far out from the slide or ram of the machine, it will spring enough to make the work a straight taper; hence keyways cut in small bores will get some of their taper from this spring.

For cutting out a half round groove, the tool shown in Fig. 170 should be employed. The outline A is made as denoted by the dotted line B in cases where, from the narrowness of the tool, it is very liable to spring from the pressure of the cut, as, say, when the thickness at C is less than three-eighths inch, in which case the cutting edge should be lowered to a straw color; whereas, if thicker, the edge may be hardened right out. It is well here to note that it is advantageous that the tool should have a barely perceptible amount of spring, in the direction of its cut, because otherwise the edge of the tool will rub against the work during the back stroke, and thus become rapidly dulled.

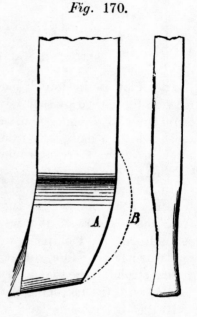

Fig. 170.

Whenever the nature of the work to be done will admit, a holding bar and short tool, such as shown in Fig. 171 may be used.

By using such a bar, short tools, such as have been already described for use in the lathe or planer, may be employed, their shortness rendering their grinding and forging much easier of accomplishment. Many of these holding bars have small pivoted boxes, similar to that shown in Fig. 171, provided to receive the tool. A is a sectional view of the bar, B is the box, pivoted at C, D is the tool, and E the set screw for holding the same. It will be observed that the set screw E screws into the pivoted box, and not into the end of the bar, and that the hole, provided in the end of the bar to admit the set screw, is large enough to permit the set screw to have plenty of play or movement. The object of this and similarly designed devices is to allow the tool to move, in the direction of D, off the pivot C, and thus to prevent the tool edge from rubbing against the sides of its cut during the up stroke of the bar, the spiral spring shown being made sufficiently strong to support the box B in the position shown, but not sufficiently strong to resist much force exerted upon the tool and in the direction of D. For small or even medium sized work these devices are

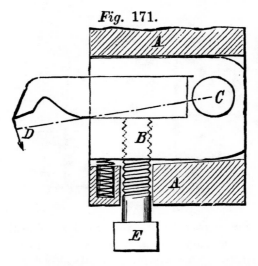

Fig. 171.

very efficient; but for large, heavy, outside work, the bars themselves are too slight, and it is usual to employ a similar device (on a large scale) provided in the tool end of the slide itself. Under these conditions the slotting machine will perform as heavy duty as either the lathe or planing machine. The writer has in his possession a cutting taken off the outside of a crank at the Morgan Iron Works, which cutting was taken at a cut $2\frac{5}{8}$ inches deep, and is a full $\frac{1}{8}$ of an inch in thickness, the tool employed being a knife tool, ground as shown in Fig. 172. B represents the tool end of the slide of the slotting machine, A the knife tool, C the work, and from D to E the depth of the cut.

The face of the tool is ground off at an angle, in the direction of I, so that the point of the tool shall not break off when it strikes the work, and so that the strain upon the tool and working parts of the machine shall not come upon them too suddenly, and cause them to break, as would be the case were the cutting edge of the tool to strike the cut along its whole length simultaneously. As

Fig. 172.

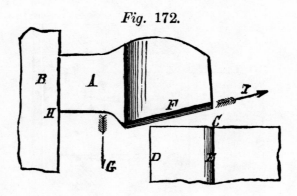

shown in the engraving, the tool would strike the work at F on the edge only, which would for an instant of time exert only enough resistance to bring all the working parts of the machine to a bearing; and as the tool descended, the strain would gradually increase until the point of the tool reached the work. When the tool is near the end of the stroke, and therefore leaves the cut, it will do so at F first, thus leaving the cut gradually, and greatly modifying the jump due to the recoil of the working parts of the machine when relieved of the heavy strain necessary to drive such a deep and thick cut. The enormous strain placed upon the tool would inevitably break it were it left very hard; it is therefore tempered to a purple.

No other tool can well be used for taking such heavy cuts, because grinding off the face, F, of any other tool would not leave the tool edge sufficiently keen to sever the metal without an excessive amount of driving power; and further, because the breadth of the face F, which sustains the force necessary to bend the cutting, is narrower in the knife tool than in any other, and therefore bends the cutting less, experiencing a corresponding decrease of strain. Cuts of such great depth and thickness cannot be well taken in slotting machines whose slides are operated by a connecting rod or link, because the excessive strain would be apt to force the connecting rod along the slot provided to alter the stroke of the machine; the sliding head is therefore provided with a strong rack on each side, operated by pinions, with suitable reversing gearing attached for varying the stroke.

When operating the feed of a slotting machine by hand, the work should be fed to the cut while the tool is reversing its motion at the top of the stroke, and not while the tool is cutting or at the bottom of the stroke, because, in either of the latter cases, the tool edge would grind against the sides of the cut during the up stroke, which would soon impair the cutting qualifications of the tool.

Tool-holding bars of sizes below about $1\frac{1}{2}$ inches in thickness should be made of steel so as to be strong enough to resist the tendency to spring. For sizes above that they may be made of wrought-iron.

CHAPTER XI.

TWIST DRILLS.

Fig. 173.

Twist drills are not, as is usually supposed, of the same diameter from end to end of the twist, but are slightly taper, diminishing towards the shank end. The taper is usually, however, so slight as to be of little consequence in actual practice. Neither are twist drills round, the diameter being eased away from a short distance behind the advance or cutting edge of the flute, backward to the next flute, so as to reduce the friction of the sides of the drill upon the hole and give the sides of the drill as much clearance as possible. The advance edges of the flutes are left of a full circle, which maintains the diameter of the drill and steadies it in the hole. If, from excessive duty, that part left a full circle should wear away at the cutting end of the drill, leaving the corner of the drill rounded, the drill must be ground sufficiently to cut away entirely the worn part, otherwise it will totally impair the value of the drill, causing it to grind against the metal, and no amount of pressure will cause it to cut. The advantage over other drills possessed by the twist drill is that the cuttings can find free egress, which effects a great saving of time, for plain drills have to be frequently withdrawn from the hole to extract the cuttings, which would jamb between the sides of the hole and the sides of the drill, and the pressure will frequently become so great as to twist or break the shank of the drill, especially in small

TWIST DRILLS.

holes. In point of fact, the advent of twist drills has rendered the employment of the flat drill for use in small holes (that is to say, from ⅜ inch downwards) totally inexcusable, except it be for metal so hard as to require a drill tempered to suit the work. Other advantages of the twist drill are, that it always runs true, requires no reforging or tempering, and, by reason of its shape, fits closely to the hole, and hence drills a straight and parallel hole, providing it is ground true.

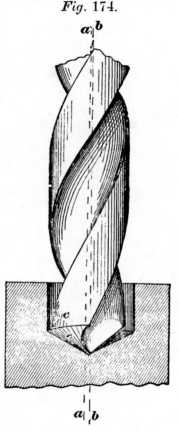

Fig. 174.

The cutting edges are usually ground to an angle of 60 degrees to the centre line of the drill, as shown in Fig. 177, but will be found to work more satisfactorily if ground to an angle of 50 degrees when used on brass work.

The line shown along the centre of the flutes, in Fig. 173, is to serve as a guide in grinding the point central when the drill is ground by hand, but more duty and more accurate work may be obtained if the drill is ground in a twist drill grinding machine, so that the cutting edges may be ground true, and both cutting edges may perform equal duty.

In Fig. 174 is shown a twist drill, having an edge e ground longer than the other, and the effect of this is that if the drill feed is $\frac{1}{100}$ inch per revolution, the whole of this feed will fall on edge *e*, and being double what it should be would cause that edge to dull more rapidly than it should do. Again, the drill would produce a hole of larger diameter than itself, because the point

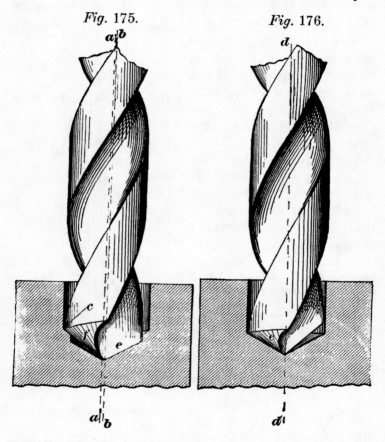

Fig. 175. *Fig.* 176.

of the drill would be forced by the feed to become the axis of revolution, which would, therefore, be on the line *b b*.

In Fig. 175 is shown a twist drill in which one cutting edge is longer than the other, and the two cutting edges are not at the same angle to the drill axis *a a*. Here again the axis of drill revolution will be on the line *b b* as before, but both cutting edges will perform duty. Thus edge *e* will pierce a hole which the outer end or corner of *f* will enlarge. Fig. 176 shows a case in which the point of the drill is central to its axis *a a*, but the

Fig. 177.   *Fig.* 178.

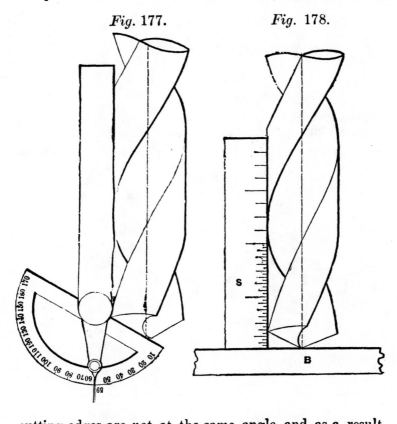

cutting edges are not at the same angle, and as a result all the duty falls on one cutting edge, and the hole drilled will be larger in diameter than the drill is, because there is a tendency for the cutting edge *e* to push or crowd the drill over to the opposite side of the hole. Drills to be ground by hand may be tested for angle in several ways. Thus, Fig. 177 represents a drill being tested by a protractor. Another method is to rest the drill upon a plate and apply a measuring rule, as in Fig. 178. Either of these methods forms a guide for testing the length of the

Fig. 179.

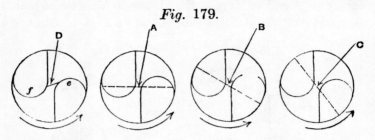

cutting edges, but they are difficult of application for very small drills, and furthermore form no guide in grinding the clearance. This, however, may be judged by the ap-

Fig. 180.

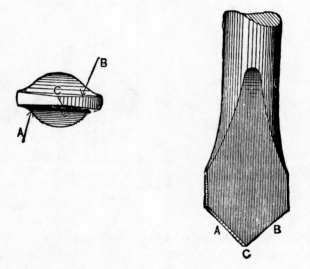

pearance of the cutting edge at that point. Thus, if this edge is at a right angle to the cutting edges, as at A, in Fig. 179, there is no clearance; while, in proportion as clearance is given, this line inclines as shown at B in the

FLAT DRILLS.

figure; at C an excessive amount of clearance is shown by the extreme angle, while at D the heels e and f are shown to stand higher than the cutting edges, because the end faces are ground at an angle in the wrong direction.

Fig. 180 represents the flat drill, whose defects are as follows: First, when the drill operates vertically, as in the drilling machine, the drill must be removed to extract the cuttings in all cases in which the depth of the drilled hole is more than about equal to its diameter, and this occupies in deep holes more time than the actual cutting operation. Secondly, the drill must be forged occasionally in order to keep it up to the required size and keep its

Fig. 181.

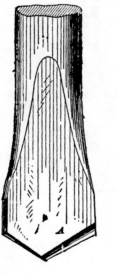

point thin enough for effective duty. Third, on account of this forging it is difficult to get the drill to run quite true, and unless this is the case the drill, when not controlled by the hole in the work, will revolve eccentrically; but so soon as the drill is subjected to end pressure on the work it will endeavor to revolve on its point as a centre of revolution, thus endeavoring to force the journal of the drill spindle to revolve eccentrically, subjecting it to undue wear and involving the loss of a large percentage of the driving power of the machine. Fourth, the diameter of the drill is dependent upon the workman's accuracy in grinding it. Fifth, the sides of the drill form but a very indifferent guide.

The keenness of a flat drill may be increased by grinding a lip to it, as at A, in Fig. 181, but a better method is to bend the lip forward when forging it, forming what is known as the lip drill. This greatly increases the

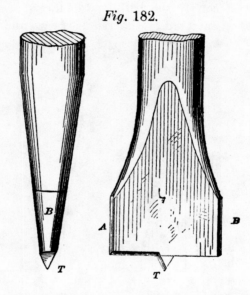

Fig. 182.

capacity of the drill for iron or copper, but causes it to rip and tear if used upon brass work.

Fig. 182 represents the tit drill, which is employed to flatten the bottoms of holes. Its tit T should be thinned at the point, as shown, and the side edges B should have little or no clearance and left so that its edges will not cut.

FEEDING DRILLS.

Much more duty may be obtained from a drill by feeding

it by hand than by permitting the gearing of the machine to feed it, because in hand feeding, the sense of feeling indicates to the operator how much cut the drill is capable of standing, and he can therefore vary the rate of feed, keeping it up to the maximum obtainable on the degree of hardness of the metal being drilled. Dullness of the cutting edges, hard or soft spots in the metal, or any other variation in the condition of the drill or in the metal being drilled, is at once perceived by hand feeding. Drilling machines have, it is true, several degrees of feed, but the fact is that the human hand can feed the drill at any rate that can be obtained by means of machine gearing; and having behind it the human mind, it is enabled to accommodate itself to the numerous and variable conditions against which no provision can be made in automatic feed gearing. No positive rate of feed, either for any size of drill, or for any particular kind of metal, can be given, because of the always present variations in the degree of hardness of the metal to be cut, and furthermore because, in the case of iron and steel, the facility of supplying the cutting edges with oil seriously affects the attainable rate of feed to the drill. If, for instance, the hole is being drilled horizontally, as in a lathe, and is very deep, so that it is difficult to freely supply the cutting end of the drill with oil, the feeding must proceed slowly or the cutting edges of the drill will soon become destroyed. Here, also, it may be well to state that, if oil be supplied to a drill cutting cast-iron or brass, it will cause the cuttings to jamb between the sides of the drill and the sides of the hole, until the pressure becomes so great as to either stop the drilling machine or lathe, or else twist or break the drill. The rate of feed, and the speed at which the drill should revolve, depend upon the hardness of the metal under operation, although not to a very great extent, except in the event of the metal being unusually hard, in which case the drill should revolve very slowly; for not much latitude in the degree of hardness of the drill is permissible, for fear of impairing the strength of the drill.

The temper for a very small flat drill should not be higher than a reddish purple, but for drills alone about ¼ inch, a coffee brown color will be found preferable. A drill may be enabled to bite very hard metal by jagging the metal surface with cold chisel marks and forcing the drill heavily to its cut, using neither oil or water, which would prevent the edges from gripping the metal.

Fig. 183.

Fig. 183 represents the Farmer lathe drill, having a straight instead of a spiral flute. This drill is accurately ground to size, and is given clearance the same as the twist drill, but the flutes are made straight along the drill instead of being spiral. By this means a keen cutting edge and free cutting are secured without the tendency to run forward when emerging through the hole, which proves so destructive to twist drills, and which is avoided. This is an important advantage upon thin work and also upon all brass work.

The shanks of machine-made drills are made parallel in the small and taper in the larger sizes, the degree of taper being $\frac{5}{8}$ inch per foot of length. Both the twist and the straight flute drill require to run at a much faster speed than the flat drill.

From some experiments made by William Sellers & Co., of Philadelphia, it has been found that a flat drill having one cutting edge at an angle of more than 104 degrees to the other will not move its position from the action of the drawing chisel.

The thinner the point of a drill, the easier it enters the metal and is fed to its cut, the limit of thinness being that which will leave sufficient strength to avoid breakage.

The angle at which to grind the end of the drill is governed to a large extent by the kind and degree of hardness of the metal to be drilled.

DRILLING HARD METALS.

Very hard metal, such as steel tempered to a blue, may be drilled by a drill tempered to a deep straw color, the drill being used at a comparatively slow speed, and forced against the work as hard as possible without breaking the point of the drill. Sufficient oil may be applied, after the point of the drill has entered the metal, to keep the cutting edges barely moist, the drill being again allowed to run dry and again moistened, thus using as small an amount of oil as is consistent with keeping the drill cool. In this way the drill will cut hard steel the best. For cast-iron, however, the drill should be kept as dry as possible. In drilling cast-iron that is very hard, and also wrought-iron that has been case-hardened, the operation may be greatly assisted by taking a hammer and a chisel, and jagging the surface of the metal, thus enabling the edges of the drill to bite it.

If necessary, the chisel may be made very hard for this especial purpose.

To make a drill exceedingly hard to suit some especial case, it may be heated in a charcoal fire to a dull red heat, and quenched in mercury instead of water. Another method is to heat the drill to a red heat in molten lead, and then to drive it into a block of cold lead, striking successive blows lightly and quickly until the drill is sufficiently cool to permit of its being held in the hand. The cases, however, in which a drill is required to be so hard are exceedingly rare.

If a drill squeaks while being operated, it arises from one of two causes: either the cutting edges are dull, and require grinding, or else the cuttings are binding in the holes. In the first case, immediate grinding is necessary; in the second, the drill should be withdrawn and the cuttings extracted. Twist drills will bring out most of the cuttings of themselves, but a piece of wire, spoon-shaped at the end, is necessary when plain drills are used.

SLOTTING OR KEYWAY DRILLS.

For drilling out oblong holes, such as keyways, or for cutting out recesses such as are required to receive short feathers in shafts, the drill known as a slotting drill, shown in Fig. 184, is brought into requisition. No. 1 is the form

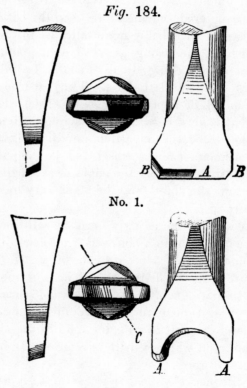

Fig. 184.

No. 1.

No. 2.

in which this tool was employed in the early days of its introduction; it is the stronger form of the two, and will take the heaviest cut. The objection to it, however, is

...hat, in cutting out deep slots, it is apt to drill out of true, the hole gradually running to one side. Suppose, for instance, as is sometimes the case, the slot or keyway is so deep that it becomes desirable to avoid having an extra long drill, which would be liable to bend and spring from the pressure of the cut, and hence that a shorter drill is used, drilling the keyway half way from each side; the tendency of such a drill would be to run to one side, so that the junction of the halves drilled from each side will not come fair.

The drill having entered on one side, and then on the other side, and having cut down until it arrived at the centre, and hence cut the keyway clear through the metal, and the junction of the two not being even at centre, it is evident that the keyway will require considerable filing to make the faces so true, level, and parallel that the key will fit all the way through. To remedy this defect, the form of drill shown in No. 2 has been brought into use. It will be observed that it enters the metal at the points A A first, and therefore cuts a ring of metal out, leaving a projecting piece in the centre which serves as a guide to steady it; whereas form No. 1 cuts a flat-bottomed hole. So that, if both drills were simply rotated and fed as a common drill, the holes made by them would appear as in Fig. 185. It

Fig. 185.

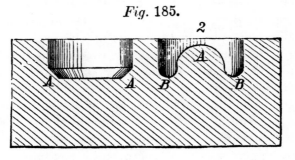

will be observed that in No. 1 the bevelled corners A alone steady the drill, while in No. 2 there is the whole core A tending to steady it, in addition to the round corners B B. In practice, however, only the round corners act to steady it, because of the light depth of the cut. These drills, are, however, never used to bore round holes, but oblong ones only, which is accomplished by either causing the drill to travel back and forth to the required length of the hole, the work being held stationary, or else by revolving the drill in a stationary position, while the table to which the work is bolted travels back and forth to the requisite distance, the cut in either case being fed to the drill at each end of the travel. Thus a slot equal to the length of the travel of the work or the drill, as the case may be, and of a width equal to the diameter of the drill, is made. If drill No. 1 is employed to cut a recess, it will leave an angular corner, while No. 2 will of course leave a round one, the bottom of the recess in either case being left quite flat, since the bottom of No. 1 is flat of itself, while the rounded corner of No. 2 cuts away as it travels along, the cone A, which, as shown in Fig. 185, is made when neither the drill nor the work travels.

Slot drill No. 1 is made by filing the cutting end square, level and true to the requisite diameter and shape, and then backing off, that is filing away on one side, the edges from the centre of the drill, outwards and across the bevelled corner, as shown in Fig. 184; while No. 2 is made by filing up the cutting end true, level, and square, and then filing out the curved hollow centrally in the end face, with a round file held at an angle with the centre line of the width of the drill, as shown by the dotted line C, in the end view of No. 2 in Fig. 184, after which the corners A A should be rounded and backed off. The thickness at the cutting end of drill No. 1 should be the same as that given for common drills, while No. 2 may be left somewhat thicker, to give it extra strength, since its form renders it comparatively weak. The reason for keeping the end of No. 1 as thin as a common drill is, that it has, at the junction of its two cutting edges, centrally on the end face and between the bevelled corners, a cutting edge across the thickness of the drill, as shown in end view, Fig. 184, and is in that respect subject to the defect before mentioned as inherent in common drills. This defect does not, however, exist in slotting drill No. 2, in which the cutting edges on the outside faces extend clear to the centre of the diameter of the drill.

Slotting drills should be tempered to a deep brown, and should be supplied freely with oil when employed to cut wrought-iron or steel, but must be kept perfectly dry when used upon cast-iron or brass. They are revolved at a higher rate of speed than common drills. To employ them in a common drilling machine whose table has no horizontal sliding motion, it is necessary to make a chuck which will bolt to the machine table; the chuck is to be provided with a pair of jaws to clamp the work, and to make the upper part of the chuck movable upon a slide in the lower part.

In using such a chuck, the operator will be very apt to vary the distance to which he moves the slide at each cut, the effect of such variation being to cause the edge of the slot or keyway to be very uneven. To remedy this, it is best, after having drilled to the proper depth, to wind the slide and set the drill so that it takes a slight cut out of one end of the slot at the top, and then (keeping the chuck stationary) to feed the drill down through the slot, thus cutting the end out quite even. In taking the first few cuts at the commencement of the operation, that is to say, immediately after the work is chucked, it is better to cut the slot a little less than the required breadth, so as to leave a little to come out of each end of the slot (as above described) to true it. It is obvious that parallel strips may be employed in the jaws, whereon to rest the work, or to make up the width between the ends of the screws, and the opposite jaw of the chuck.

There is probably no one cutting tool used in a machine which saves so much labor as the slotting drill, because it performs a duty that no other tool or machine can perform, and which is moreover a most difficult and tedious one. Before the advent of this tool, deep keyways were cut out of the solid metal in the following manner: first, plain holes were drilled through the work, then these holes were plugged up by having pieces of round iron driven tightly in them. Then new holes were drilled, the centre of each new hole being in the thin wall of solid metal between the plugged holes. After the latter holes were drilled, the remains of the plugs were driven out, the sides of the keyway would present a serrated appearance. This entailed an almost incredible amount of chipping and filing in order to make the sides of the keyway level and true, and the width parallel. This method of procedure is, however, still in vogue to a slight extent, being confined mainly to jobbing and repair shops. It is also employed for very narrow and deep holes, since a slotting drill cannot be employed to advantage in holes of less than about $1\frac{5}{8}$ of an inch in diameter, because of the bending and springing of the drill. If, however, twist drills are employed to drill the small holes, the plugging with pieces of iron may be dispensed with, since the holes may be drilled so as to run one into the other; in this case every other hole should be drilled first, and then in drilling the intermediate holes the drill will not run to one side or the other.

It may here be observed that the principles of the action of the slot drill have been applied to a variety of purposes in woodworking, prominent among which is its use in Boult's panelling and dovetailing machine. In its adaptation to wood, as in its adaptation to iron, there is no other tool at all capable of performing the same kind of duty, irrespective of either time or quality.

Keyways or slots that are wide enough to admit a stout

PIN DRILLS.

tool may be cut by drilling a hole the width of the required keyway at one end, and then cutting out the remainder of the keyway in the slotting machine. All ordinary keyways, however, are cut quicker and better with the slot drill.

PIN DRILLS.

The next form in which the drill appears is the pin drill, which is a drill having a pin projecting beyond and between its cutting edges, as shown in Fig. 186, A A being the cut-

Fig. 186.

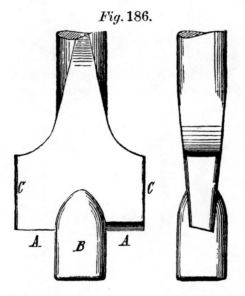

ting edges. The purpose of this drill is to face off the metal round the outside of holes, the pin B fitting into the hole so as to steady the drill and keep it true with the hole. In making this tool, the pin B, the edges C, and the ends forming the cutting edges A A, should be turned up true in the lathe; the backing off may then be filed, leaving the cutting edges A with the turning marks barely effaced; thus they will be sure to be true and at an equal height from the end of the pin, so that both the cutting edges will perform duty, and not one only, as would be otherwise the case. Pin drills should be tempered to a deep straw color, and run at a comparatively slow speed, using oil for wrought-iron and steel, and running dry on cast-iron and brass. In cases where, for want of an assortment of pin drills, there is none at hand with a pin suitable for the size of hole required to be faced, a drill having one too small for the hole may be made up to the required size by placing upon it a ring of iron or brass of the requisite thickness and about equal in depth to the pin.

COUNTERSINK DRILLS.

Of countersinks, there are various forms; but before proceeding to describe them, it may be as well to observe that the pin drill described above may be employed as a flat-bottomed countersink. Fig. 187 represents a taper counter-

Fig. 187.

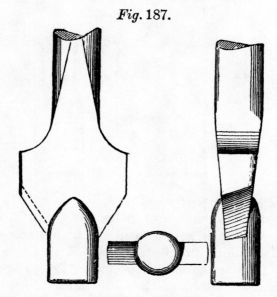

sink, such as is employed for holes to receive flush rivets or countersunk head bolts, this form of tool being mainly employed for holes above $\frac{5}{16}$ of an inch in diameter, A A

COUNTERSINKS.

being in each case the cutting edge, and B the pin. It should be made, tempered, and used as directed for pin drills. In tempering these tools, or any others having a pin or projection to serve as a guide in a hole, the tool should be hardened right out from the end of the pin to about ⅜ of an inch above the cutting edges. Then lower the temper of the metal (most at and near the cutting edges), leaving the pin of a light straw color, which may be accomplished by pouring a little oil upon it during the lowering or tempering process. The object of this is to preserve it as much as possible from the wear due to its friction against the sides of the hole. For use on wrought-iron and steel, this countersink (as also the pin drill) may have the front face hollowed out, after the fashion of the lip drill.

For use on holes ½ inch and less in diameter, we may use a countersink made by turning up a cone, and filing upon it teeth similar to those upon a reamer, or we may take the same turned cone and cut it away to half its diameter, similar to a half round bitt. Either of these countersinks will cut true and smoothly, oil being applied when they are used upon steel or wrought-iron.

Common drills, ground to the requisite angle or cone, are sometimes used as countersinks, but they are apt to cut

Fig. 188.

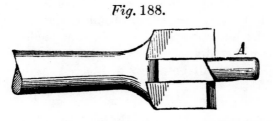

untrue and uneven. For fine and light work, the pin drill, with its cutting edges either at right angles to the centre line of the pin or at such other angle as may be required, forms the best countersink; it should, however, have more than two cutting edges, so that they may steady it. Fig. 188 presents an excellent form of this tool, A being one of the four cutting edges.

This tool is formed by turning up the whole body, filing out the necessary four spaces between the cutters, and backing the latter off at the ends only, so that the circumferential edges will not cut, and hence the recesses or countersinks will be all of one diameter. When to be used upon steel or iron only and not upon brass, the flutes may be given angle, as in Fig. 189, which will enable the tool to cut more freely. It is obvious that the capacity of tools of this class may be increased by making the pin A as small as is compatible with strength, and then fitting thereto small rings or bushes, so as to vary the diameter of the pin to suit the size of the hole. Or in the larger sizes the pin may be let into a taper hole so that various sizes of solid pins may be used. It is necessary, however, to the production of round holes that shall be of uniform diameter that the pins fit to the hole without any play or lost motion, so that the tool shall not have liberty to move. In cases of emergency this end may be served by a ferule made of sheet tin or brass merely bent around upon the pin. To ensure the accurate fit of the pin to its hole it may be given cutting edges, and to facilitate the sharpening of these edges a small hole may be pierced up the centre of the pin.

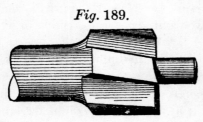

Fig. 189.

CUTTERS.

Cutters are steel bits, usually held in either a stock or bar, being fitted and keyed to the same; by this means cutters of various shapes and sizes may be made to fit one stock or bar, thus obviating the necessity of having a

CUTTERS.

multiplicity of these tools. The end served by the use of cutters is to avoid the necessity of cutting away all the metal in order to produce the hole, it being obvious that a cutter makes an annular groove leaving a ferule, which falls out when the cutter has been fed entirely through the work. The width of the cutter edge or groove is not usually more than about $\frac{5}{16}$ inch. In place of a plain, a tubular cutter is sometimes used. Such cutters, however, are expensive to make and involve more trouble to grind for the resharpening, on which account a single or a pair of straight cutters are generally preferred. Of cutter stocks, which are usually employed to cut holes of comparatively large diameter, as in the case of tube plates for boilers, there are two kinds, the simplest and easiest to be made being that shown in Fig. 190.

A is the stock, through which runs the slot or keyway into which the cutter B fits, being locked by the key C. D is a pin to steady the tool while it is in operation. Holes of the size of the pin D are first drilled in the work, into which the pin fits. To obviate the necessity of drilling these holes, some modern drill stocks have, in place of the pin D, a conical-ended pin which acts as a centre, and which fits into a centre punch mark made in the centre of the hole to be cut in the work.

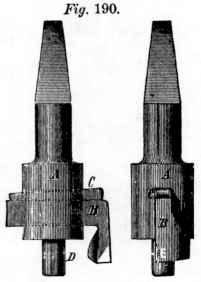

Fig. 190.

Most of these devices are patented, and the principle upon which they act will be understood from Fig. 190a, A being the stock to which the cutters B B are bolted with one or more screws. C is a spiral spring working in a hole in the stock to receive it. Into the outer end of this hole fits, at a working fit, the centre D, which is prevented from being forced out (from the pressure of the spring C) by the pin working in the recess, as shown. E is the plate to be cut out, from which it will be observed that the centre D is forced into the centre punch mark in the plate by the spring C, and thus serves as a guide to steady the cutters and cause them to revolve in a true circle, so that the necessity of first drilling a hole, as required in the employment of the form of stock shown in Fig. 58, is obviated. The cutters are broadest at the cutting edge, which is necessary to give the point clearance in the groove. They are also at the taper part (that is to say, the part projecting below the stock) made thinner behind than at the cutting edge, which is done to give the sides clearance. It is obvious that, with suitable cutters, various sized holes may be cut with one stock.

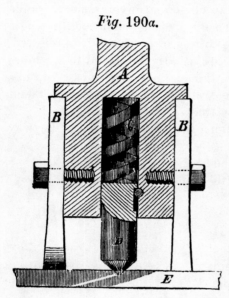

Fig. 190a.

In cutting out holes of a large diameter in sheet-iron, a stock and cutter, such as shown in Fig. 191, is generally employed; but the great distance of the cutting edge from the stock centre, that is to say, the extreme length of the cutter bar, renders it very liable to spring, in which case these and other tools having a slight body and broad cutting edge are almost sure to break, unless some provision is made so that the tool, in springing, will recede from and not advance into the cut. To accomplish this end, we must shape the cutter as shown in Fig. 191, which will, at the very least, double the efficiency of the tool.

In Fig. 191 the cutting edge B stands in the rear of the line A, or fulcrum from which the springing takes place; hence, when the tool springs, it will recede from the work C. To avoid springing and for very large holes, the cutter may be a short tool, held by a stout crossbar carried by the stock; but in any event the cutter should be made as shown above.

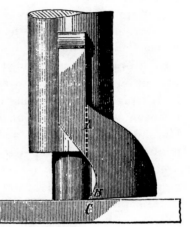

Cutters of a standard size, and intended to fit the pin stock, should be recessed to fit the end of the slot in the stock. In making these cutters they should be first fitted to the stock, and then turned up in the lathe, using the stock as a mandril, the ends being then backed off to form the cutting edges. Those slight in substance should be tempered to a light straw at the cutting edge, and left softer at the back part. Those above five-sixteenths of an inch in thickness may be hardened right out and not tempered at all.

Here it may be as well to describe a process for tempering cutters, which, as several very expert workmen have assured me, gives superior results. It is to heat the cutter to a cherry-red heat, and quench it in water until it is cold, and to then reheat it until water dropped upon it will dry off in slight bubbles. If, however, the reheating is rapidly performed, there will be no need to drop any water on it, since that which adheres to it after quenching will be sufficient. I have no doubt but that for stout cutters, or even for slight ones which perform a light duty, this method is preferable to all others: but for light cutters performing a heavy duty, I should judge that it would leave them too hard for their strength, and therefore liable to break.

Cutters for boring bars should be, if intended to be of standard size, recessed to fit the bar, the bar having a flat place filed around and beyond the edges of the hole, to form a broader bearing for the cutter to fit upon. But, if the cutter is intended to vary the size of hole, it must be left plain, so that it may be moved inwards or outwards to accommodate the size of bore required. All cutters and bits should be used at a cutting speed of about 15 feet per minute, and with oil or soapy water for work in wrought-iron or steel; and for use on those metals, the cutters, etc., may be given a little front rake by grinding away the metal of the front face, as shown by the dotted line in Fig. 190, at E.

CHAPTER XII.

TOOL STEEL.

The cutting tools for all machines should be made of hammered (which is tougher and of finer grain than rolled) steel. Even in a bar of hammered steel, the corners, from receiving the most effect from the action of the hammer, are of better quality (that is, more refined) than the rest of the bar. This fact is clearly demonstrated in the manufacture of the celebrated Damascus swords and gun barrels, in which the square bars of metal are, after being hammered, twisted and then hammered square again; the twisting process is then repeated, and the bar again forged square, the whole operation being repeated until the body of the entire bar is completely intersected with metal which has, at some time during the forging process, formed the corners of a square. The effect of this treatment becomes apparent upon immersing the metal in acid, which will eat away those parts which have not formed a corner at some stage of the process of manufacture, more rapidly than the rest of the metal, and that to such a degree as to give to the whole the appearance of having been engraved, thus evidencing that the parts that have received the most hammering are of finer quality than the rest of the bar.

For cutting tools, it is highly necessary to gain every attainable superiority in the steel; and if we cannot take three months of time to prepare bars for this special purpose (as they do in the above process), we can at least employ well-hammered steel, and thus secure the best known practicable results.

The test of tool steel is the speed at which it will cut and the length of time it will last without being ground, concerning which it is difficult to get data, unless by actual experiment with different kinds of steel upon work of the same diameter and texture of metal, because the cutting speed employed by workmen varies as much as 10 feet per minute upon the same diameter of work. The use of more than one kind of tool steel in a workshop should always be avoided, because different kinds of steel require different treatment, both in forging and hardening; and when more than one kind is in use in the shop, the whole of them are liable (from not noticing the particular brand) to wrong treatment.

Mushet's "special tool steel" makes an excellent tool for roughing work out on the lathe or planer, and will undoubtedly stand a higher rate of cutting speed than other steel. For turning shafting and cast-iron pulleys, especially when the latter are unusually hard, this tool works admirably. Its peculiarity is that it is hard of itself, and therefore requires no hardening. Immersing it in water when it is heated causes it to crack. The advantages claimed for it are its high rate of cutting speed, and that it is easily ground, since it will not soften by heating during the operation. It is, on the other hand, difficult to forge in consequence of its excessive hardness even when heated; it must not be forged at so great or so low a temperature as other steel, or it will crack; and as it is not adapted for general tool purposes, its disadvantages, independent of its increased cost, render its introduction into the *general* machine-shop unadvisable.

FORGING TOOLS.

The forging of a tool should be formed in as few heats as possible, for steel deteriorates by repeated heating, unless it is well hammered at each heat; and if the tool has a narrow edge, care should also be taken to hammer it on that edge before the metal has lost much of its heat.

and to strike it more lightly as it gets cooler, for striking a narrow surface of steel when it is somewhat cool has the same injurious effect upon it as striking it endwise of the grain (which is termed upsetting it), destroying its cutting value and strength.

If the tool requires much drawing out, the steel should be drawn rectangular to as near as possible the required size, leaving only sufficient metal to shape up the tool. The steel should be heated to a light-red and in no case to a yellow heat, which would cause it to become what is termed "*burnt*," after which its cutting value is irretrievably lost, nor will any amount of forging restore it to its former standard. A tool that has been accidentally overheated should have the burnt part cut off and be entirely reforged. Front or other tools for use on wrought-iron should be forged as follows: Draw the bar steel down taper on one edge of the steel and parallel on the other, and then bend the cutting part of the tool up so that the tool will bear the shape of Figs. 4, 5, 6, and 7, which are better and more easily forged tools than those shown by Figs. 9, 19, and 18, the latter being *fac similes* of the forms of tools in use at the Morgan Iron Works. They are of perfectly correct form so far as the shapes of the cutting edges are concerned, but are more difficult to forge and to grind than the former, which are again superior in that the height of their cutting edges from the bottom edge is less, and hence they do not suffer so much from the causes explained by Figs. 24 and 25 and their accompanying remarks. It is obvious, however, that the height of the tool edge from the bottom face of a lathe tool is determined by the height of the face of the slide rest on which the tool rests from the horizontal centre of the lathe centres.

In using American chrome steel, be careful to forge it according to the directions supplied by its manufacturers, its treatment being almost the opposite for that applicable to English tool steel, the former requiring to be heated to a much higher temperature for forging, and to a less temperature for hardening, than the latter.

TOOL HARDENING AND TEMPERING.

Steel is said to be hardened when it is as hard as it is practicable to make it, and to be tempered when, after having been hardened, it is subjected to a less degree of heat, which partly but not altogether destroys or removes the hardness. The degree to which this tempering is performed, or in other words the degree of the temper, is made perceptible and estimated as follows: By heating a piece of steel to a red heat (not so hot as to cause it to scale), and then plunging it into cold water and allowing it to remain there until it is cold, it will be hardened right out, as it is termed, that is, it will be made hard to the greatest practicable degree. If it is then slowly reheated, its outer surface will, as the temperature increases, assume various shades of color, commencing with a very light straw color, which deepens successively to a deep yellow, red, brown, purple, blue, and green, which latter fades away as the steel becomes heated to redness again, when the effects of the first hardening will have been entirely removed. It becomes apparent, then, that the colors which appear upon the surface of the steel denote the degree to which the tempering or resoftening operation has taken place. Having then by practice ascertained the color which denotes the particular degree of hardness requisite for any specified tool, we are enabled to always temper it to that degree, sufficiently near for all practical purposes. It is undoubtedly true that, if the conditions of tempering which will be laid down in all our instructions are (for want of sufficient experience in the operator) varied, the colors will not present, to positive exactitude, the precise degree of temper: the difference being that, if the color forms very rapidly, the tool may be left of a lighter color; and that if the colors form very slowly, the tool may be left of a slightly deeper hue. The difference in temper, however, as compared to the color, will in no case be sufficient to be perceptible in ordinary tool practice, and need not, save under circumstances requiring great minuteness in the degree of temper, be paid any attention to.

When a tool such as a drill requires to be tempered at and near the cutting edge only, and it is desirable to leave the other part or parts soft, the tempering is performed by heating the steel for some little distance back from the cutting edge, and then immersing the cutting edge and about one-half of the rest of the steel, which is heated to as high a degree as a red heat, in the water until it is cold; then withdraw the tool and brighten the surface which has been immersed by rubbing it with a piece of soft stone (such as a piece of a worn-out grindstone) or a piece of coarse emery cloth, the object of brightening the surface being to cause the colors to show themselves distinctly. The instant this operation has been performed the brightened surface should be lightly brushed by switching the finger rapidly over it; for unless this is done, the colors appearing will be false colors, as will be found by neglecting this latter operation, in which case the steel after quenching will be of one color; and if then wiped, will appear of a different hue. A piece of waste or other material may of course be used in place of the hand. The heat of that part of the tool which has not been immersed will become imparted to that part which was hardened, and, by the deepening of the colors, denote the point of time at which it is necessary to again immerse the tool and quench it altogether cold.

The operation of the first dipping requires some little judgment and care; for if the tool is dipped a certain distance and held in that position without being moved till the end dipped is cold, and the tempering process is proceeded with, the colors from yellow to green will appear in a narrow band, and it will be impossible to directly perceive when the cutting edge is at the exact shade of color required; then again, the breadth of metal of any one degree of color will be so small that once grinding the tool will remove it and give us a cutting edge having a different degree of temper or of hardness. The first dipping should be performed thus: Lower the tool vertically into the water to about one-third of the distance to which it is red hot, hold it still for about sufficient time to cool the end immersed, then suddenly plunge it another third of the distance to which it is heated red, and withdraw it before it has had time to become more than half cooled. By this means the body of metal between the cutting edge and the part behind, which is still red hot, will be sufficiently long to cause the variation in the temperature of the tool end to be extended in a broad band, so that the band of yellow will extend some little distance before it deepens into a red; hence it will be easy to ascertain when the precise degree of color and of temper is obtained, when the tool may be entirely quenched. A further advantage to the credit of this plan of dipping is that the required degree of hardness will vary but very little in consequence of grinding the tool; and if the operation is carefully performed, the tool can be so tempered that, by the time the tool has lost the required degree of temper from being ground back, it will also require reforging or reforming. As a rule a tool should be made to a red heat to a distance about twice the diameter of the tool steel of which it is made.

The degree to which a tool may be hardened is dependent in a great measure upon its shape. The only reason for tempering any lathe tool is to strengthen it, for steel hardened right out is comparatively weak and gains strength by being tempered. The lower the temper the greater the strength. A straw color is well adapted to ordinary light tools, but very slight tools, such as say a parting tool $\frac{1}{16}$

nch wide, may be lowered to a deep brown or almost to a purple. Stout tools, such as are shown in Fig. 2, may be made as hard as fire and water will make them; so also may the tools presented in Figs. 8, 9, 18, 19, 28, and 34; while slight tools, such as are given in Figs. 29 and 30, should be lowered in temper to a light straw color.

The practice of lowering stout tools to a straw color is sometimes resorted to, but it is certainly an error, for it is undoubtedly advantageous to make the tool as hard as it can be made, so long as it will bear the strain of the cut, which is possible and easy of accomplishment with Jessop's, Moss', Sanderson's, or other similar grades of tool steel.

If a tool so hardened is found to break, it is in consequence either of its being bad steel or else it has been heated to too great a temperature in the process of forging or hardening, unless it has been given too much rake for the duty to which it has been allotted. Tool steel may be forged at such a temperature that it is not positively burned, and yet has lost part of its virtue; and while under such circumstances it would break if hardened right out, it will cut and stand moderately well if the temper be lowered to a straw color.

This is simply sacrificing the degree of hardness to cover the blunder committed by overheating, and it is from such causes that the variation of cutting speed employed by mechanics arises; for a youth who has learned his trade in a shop where the tools were overheated, and consequently underhardened, settles down to the rate of cutting speed attainable under those circumstances and adheres to it; while he who has been accustomed to the use of tools properly forged and hardened right out, upon entering another shop where the tools are overheated in forging and underhardened to compensate for it, finding he cannot get the cutting speed up to his customary rate, breaks off the tool point to see if it has been burned, and, finding that the grain of the metal does not appear granulated, sparkling, and coarse, as it would do if positively burned, condemns the quality of the steel.

The grain of properly forged and hardened tool steel appears, when fractured, close and fine, and of a dull, whitish tint, the fracture being even on its surface.

American chrome tool steel may be made unusually hard by using very clean water and adding a piece of fuller's earth and a piece of common soda, each of the size of a hazel nut, to a pailful of water.

In all cases where a tool can be ground to sharpen it, it should be hardened before grinding, for steel hardened with the forged skin on is stronger and better than that in which the skin is removed before hardening. When it is intended to harden a tool right out, heat it to a cherry red to the distance that it is necessary to harden it, and plunge it into the water suddenly to the distance it requires hardening; hold it still a moment, then dip it a little deeper, and withdraw it again to the amount of the last dipping, repeating this latter operation until the tool is cold; for by this means the junction of the hard and soft steel in the tool is graduated and not sharply defined, the result being that the tool is less liable to fracture either in hardening or in using. If the tool to be hardened has a thick part to it, let that part enter the water first and immerse the tool slowly, so that it will be cooled as nearly equally as possible and thus be prevented from cracking in hardening.

Tools heated by charcoal are much superior to those heated by common coal, and need not be made quite so hot to harden. To harden steel, never get it hot enough to cause it to scale. Thin pieces of steel, and taps, dies reamers, drifts, and similarly shaped tools, should be dipped endways; for if dipped otherwise, they are sure to warp in hardening. Very slight tools may be prevented from cracking by making the water quite warm before immersing them and then holding them still in the water; in fact, alwater for hardening purposes should have the chill off it by heating, before being used, or the articles hardened in it are very liable to crack. If the article requires to be hardened all over, immerse it (suspended on a wire hook) so that the water may have free and equal access to the whole surface of the steel, which is not possible with tongs in consequence of their jaws covering part of the steel.

HARDENING.

All work to be hardened should be heated according to its shape, the work being so manipulated in the fire that the thin parts do not get to the required heat before the thick parts do. Then in quenching them in the water the thick parts should be immersed first, and the operation be performed slowly. The work should be lowered perpendicularly in the water and immersed deeply, and not under any circumstances moved sideways. Uneven heating warps the work in the fire, careless dipping warps and cracks it in the hardening. Always use water that is at least lukewarm, and if the article has one part much thinner than another, or is very slight, and hence liable to warp or crack, make the water as hot as the hand will bear it, and dip the work edgeways, the heaviest side being downwards. Very small articles to be hardened in quantities may be heated in a piece of wrought-iron pipe, having one end closed, the pipe being revolved in the fire during the heating process to equalize the heating of the work.

TO HARDEN SPRINGS.

Small springs, which should be made of spring-steel or double-shear steel, may be hardened as follows: Heat them to a bright cherry-red, and quench them in water having the cold chill taken off it. If, on being taken from the water, they are white, or mottled with white and a light gray, they are hard enough, but if they are dark-colored, they are not hard enough, and must be rehardened. After being hardened, they may be tempered as follows: String them, if possible, on a wire, and fry them over the fire in a pan or tray containing enough lard oil to well cover them, and heat the oil until it will blaze all over the surface, then turn the springs over and over in the blazing oil, letting them blaze long enough to be sure that the thick parts of the spring are equally heated with the thin parts. If a single spring requires tempering, it may be tempered by fastening it to a wire, and just above it put a small roll of wire to retain the oil. Heat the spring over a very slow fire, and apply oil, letting it run down the wire to the spring. Keep the spring supplied with oil, and let it blaze a minute or so. If it has a light or thin part to it, pour cold oil on that part of it during the early part of the blazing process.

Large springs are first hardened, and then blazed off in whale oil, containing 2 lbs. of tallow and ¼ lb. of beeswax (or, instead of the latter, ½ lb. of black resin) to every gallon of whale oil. If a spring is made of cast-steel it must, after blazing off, be left to cool of itself without being quenched off.

Springs that have the forged skin on are stronger and more elastic than those which are brightened, and all springs are reduced in elasticity by grinding off the surface after they are tempered; especially, however, is this the case with those having the forged or rolled skin on.

To harden machine-steel, or make cast-steel very hard, put a pound of salt to a gallon of rain-water.

The longer water or a tempering liquid is used the better it becomes, but either of them are wholly spoiled if any greasy substance gets in them.

All steel, as well as iron, swells by hardening, so that holes become smaller, and outside surfaces larger, in consequence of hardening, and this fact is often taken advantage of to refit iron or steel work that has become worn. For instance, suppose a bolt has worn loose: the bolt may be hardened by the common prussiate of potash process, which will cause it to increase in size, both in length and diameter. The hole may be also hardened in the same way, which will decrease its diameter; and if the decrease is more than necessary, the hole may be ground or "lapped" out by means of a lap. Only about $\frac{1}{84}$ of an inch of shrinkage can be obtained on a hole and bolt by hardening, which, however, is highly advantageous when it is sufficient, because both the hole and the bolt will wear longer for being hardened.

CASE-HARDENING WROUGHT-IRON.

Iron may be case-hardened, that is, the surface converted into steel and hardened, as follows: First, by the common prussiate of potash process, which is as follows: Crush the potash to a powder, being careful that there are no lumps left in it, then heat the iron as hot as possible without causing it to scale; and with a piece of rod iron, spoon-shaped at the end, apply the prussiate of potash to the surface of the iron, rub it with the spoon end of the rod until it fuses and runs all over the article, which must then be placed in the fire again and slightly reheated, and then plunged into water, observing the rules given for immersing steel so as not to warp the article.

Another method is to place the pieces to be hardened in an iron box, made air-tight by having all its seams covered well with fire clay, filling the box in with bone dust closely packed around the articles, or (what is better) with leather and hoofs cut into pieces about an inch in size, adding thin layers of salt in the proportion of about 4 lbs. salt to 20 lbs. of leather and 15 lbs. of hoofs. In packing the articles in the box, be careful to so place them that when the hoofs, leather, etc., are burned away, and the pieces of iron in the box receive the weight of those above them, they will not be likely to bend from the pressure. When the articles are packed and the box ready to be closed with the lid, pour into it one gallon of urine to the above quantities of leather, etc.; then fasten down the lid and seal the seams outside well with clay. The box is then placed in a furnace and allowed to remain there for about 12 hours, when the articles are taken out and quickly immersed in water, care being taken to put them in the water endways to avoid warping them.

Articles to be case-hardened in the above manner should have pieces of sheet-iron fitted in them in all parts where they are required to fit well and are difficult to bend when cold, and the heaviest pieces of work should be put at the bottom of the box.

THE WEAR OF METAL SURFACES.

The wear of metal surfaces, such as cast-iron, wrought-iron, steel, and brass, is governed as much by the conditions under which that wear takes place as it is by the degree of hardness of the metal.

It is a general rule, that motion in one continuous direction causes more wear, under equal conditions, than does a reciprocating motion, and also that the harder the metal the less the wear. To this latter rule there are, however, exceptions in favor of cast-iron, which will wear better when surrounded by steam than will any other metal. Thus, for instance, experience has demonstrated that piston-rings of cast-iron will wear smoother, better, and equally as long as those of steel, and longer than those of either wrought-iron or brass, whether the cylinder in which it works be composed of brass, steel, wrought-iron, or cast-iron—the latter being the more noteworthy, since two surfaces of the same metal do not, as a rule, wear or work well together. So also slide-valves of brass are not found to wear so long or so smoothly as those of cast-iron, let the metal of which the seating is composed be whatever it may; while, on the other hand, a cast-iron slide-valve will wear longer of itself, and cause less wear to its seat, if the latter is of cast-iron, than if of steel, wrought-iron, or brass. The duty in each of these cases is light; the pressure on the cast-iron, in the first instance cited, probably never exceeding a pressure of ten pounds per inch, while, in the latter case, two hundred pounds per square inch of area is probably the extreme limit under which slide-valves work; and what the result under much heavier pressures would be is entirely problematical.

Cast-iron in bearings or boxes is found to work exceedingly smoothly and well under light duty, provided the lubrication is perfect and the surfaces can be kept practically free from grit and dust. The reason of this is, that cast-iron, especially that of American manufacture, forms a hard surface-skin, when rubbed under a light pressure, and so long as the pressure is not sufficient to abrade this hard skin, it will wear bright and very smooth, becoming so hard that a scraper made as hard as fire and water will make it will scarcely cut the skin referred to. Thus, in making cast-iron and wrought-iron surface-plates or planometers, we may rub two such plates of cast-iron together under moderate pressure for an indefinite length of time, and the tops of the scraper-marks will become bright and smooth, but will not wear off; while if we rub one of cast-iron and one of wrought-iron, or two of wrought-iron, well together, the wrought-iron surfaces will abrade so that the protruding scraper-marks will entirely disappear, while the slight amount of lubrication placed between such surfaces to prevent them from cutting will become, in consequence of the presence of the wrought-iron, thick and of a dark-blue color, and will cling to the surfaces, so that after a time it becomes difficult to move the one surface upon the other. If, however, the surfaces are pressed together sufficiently to abrade the hard skin from the cast-iron, a rapid cutting immediately takes place, which is very difficult to remove, the only remedy being to entirely remove the particles of metal due to the abrasion, and lubricate very freely.

Under a light duty, cast-iron, especially when working under steam pressure, will wear longer and better than brass, wrought-iron or steel, even if the motion be continuously in one direction; thus, for revolving side-surfaces, such as discs, it retains its superiority over the harder metals, and there is no test so great as is involved under such conditions, for the following reasons:

Suppose we have a piston revolving in a cylinder. The metal on the piston, at a distance of 2 inches from its centre, will pass over a circle, in the cylinder, of 12,566 inches in circumference. The metal on the piston, however, at a distance of 4 inches from its centre, will pass over a circle or surface of a circumference of 25,132 inches. Thus, we find the one part of the piston to pass over twice as much metal as does the other in performing a revolution, making the wear on *that* account twice as great at the large radius as it is at the small one. But this is not all, for the metal at the large radius travelled over its wearing surface, that is to say, the surface it bears against, in making a revolution, at a speed twice as great as did the metal at the small one over its wearing surface, since one travelled over sixteen inches in the same time that the other travelled over eight inches of surface; this increase further doubles the wear at the large radius, making its wear fourfold that at the small one, and giving us the rule that the wear of a revolving disc increases (as does its area) in the ratio of the square of its diameter. The result of this inequality of wear was demonstrated in the early days of locomotive-engineering, at which time the throttle-valves were in nearly all engines semi-revolving discs, with radial openings, the wearing surfaces being on the side face, and the disc revolving reciprocally on a centre-pin.

The result of the wear on such valves was found to be very unsatisfactory, because the metal at and near the extreme circumference would wear very rapidly away. The pressure of the steam, however, by springing the outer surface of the disc to its seat, would prevent the faces from leaking, but the pressure of the outer diametrical surface to its seat would be diminished in proportion to the resistance of the metal to the spring referred to, and, as a consequence, the surface of the metal at and near the centre of the disc would have upon its bearing surface not only the pressure due to the steam acting upon its exposed surface, but an amount in excess equal to that to which the outer diameter was relieved in consequence of its resistance to spring.

These conditions would continue until the wear of the larger diameter becoming greater, and the amount of spring required to keep it to its seat increasing in proportion, the resistance of the metal to so much spring partly relieves the pressure of the larger diameter to its seat, and since the pressure due to the force exerted by the steam upon the exposed surface of the disc will remain constant, to whatever amount the outer diameter is relieved of the pressure to its seat, that at and near the centre, forcing it to its seat, will be augmented, until at last the excessive pressure will cause it to cut or abrade, which action will continue until the cutting at and about the centre will allow the larger diameter to bed with more force to its seat, by diminishing the amount of its spring, and hence its resistance to the steam-pressure immediately behind it, whereupon its excessive wear would recommence.

If, however, the thickness of the disc were made such as to enable it to resist the steam-pressure without springing, the larger diameter would wear sufficiently away to cause the valve to leak; whereas, if the disc were made sufficiently thin to enable it to spring easily, the outer diameter would wear to almost a feather-edge, while the metal about the centre would nearly maintain its original thickness.

It is this inequality of wear in revolving, or side, or disc surfaces that is the stumbling-block to the success of rotary engines, nor has there as yet been suggested any method of overcoming or compensating for it. It is difficult, indeed, to perceive in what direction such a remedy can lay, unless it be in making the disc of hardened steel and tempering it, so that being at the outer diameter as hard as fire and water will make it, it is so tempered that it shall be gradually softer as the diameter decreases, until at the centre it is quite soft. Thus the degree of hardness of the metal will be as far as possible in proportion to its liability to wear.

In an experiment made by me, I revolved two cast-iron disc-surfaces, of three inches diameter, under a pressure of steam of 20, 35, and 70 lbs. alternately, per square inch, the surface being pressed together under a pressure of about 7 lbs. per square inch, and the discs making three thousand revolutions per minute. I found that, in consequence of the light pressure, forcing the faces together, a cast-iron surface showed but very little signs of wear—not sufficient, indeed, after running ten hours a day for ten days, to efface the scraper-marks from the surfaces, which had become polished and glazed, as it were. Several small holes were then drilled in the contacting surfaces, and plugs of Babbitt metal, brass, wrought-iron, and steel, were inserted, the faces being rescraped all over, and the discs then run as before, the result being that, after two days of running, the cast-iron appeared smooth and bright as before, while the brass, wrought-iron, steel, and Babbitt metal were found to be worn positively below the surface of the cast-iron, several repetitions of the last experiment giving, in each case, a like result.

The reason that the liability to cut is found in practice to be much greater in revolving than in reciprocating surfaces is that, when a revolving surface commences to cut, the particles of metal being cut are forced into and add themselves, in a great measure, to the particles performing the cutting, increasing its size and the strain of contact of the surfaces, causing them to cut deeper and deeper until at least an entire revolution has been made, when the severed particles of metal release themselves, and are for the most part forced into the grooves made by the cutting.

In reciprocating surfaces, when any part commences to cut, the edge of the protruding cutting part is abraded by the return stroke, which fact is clearly demonstrated in either fitting or grinding in the plugs of cocks, in which operation it is found absolutely necessary to revolve the plugs back and forth, to prevent the cutting which inevitably and invariably takes place if the plug is revolved in a continuous direction. Furthermore, when a surface revolves in a continuous direction, any grit that may lodge in a speck, hollow spot, or soft place in the metal, will cut a groove and not easily work its way out, as is demonstrated in polishing work in a lathe; for be the polishing material as fine as it may, it will not polish so smoothly unless kept in rapid motion back and forth. Grain emery used upon a side-face, such as the outer face of a cylinder-cover, will lodge in any small, hollow spots in the metal and cut grooves, unless the polishing stick be moved rapidly back and forth between the centre and the outer diameter. If a revolving surface abrades so much as to seize and come to a standstill, it will be found very difficult to force it forward, while it will be comparatively easy to move it backward, which will not only release the particles of metal already severed from the main body, and permit them to lodge in the grooves due to the cutting, but will also dislodge the projecting particles which are performing the cutting, so that a few reciprocating movements and ample lubrication will, in most cases, stop the cutting and wash out the particles already cut from the surfaces of the metal.

It is held by many that fast-running bearings filled with Babbitt metal will wear better than brass bearings. Such, however, is not the case, if the bearings are properly fitted; the only advantage possessed by Babbitt metal bearings is that they are more easily fitted; because the Babbitt will run so as to make an even and equal bearing upon the shaft, and it is therefore only necessary to set the shaft true before pouring the metal, to insure an even and true bearing; whereas, after the brasses are fitted and bored, they require fitting to the shaft while in their places, and this being a somewhat tedious operation is often omitted, the consequence being that the journal does not bed fairly on the bearing surface, and thus the whole strain of the bearing is placed upon the reduced surface of the brasses which beds upon the journals, and the increase of journal-pressure per inch, placed upon the brass, induces undue abrasion, and a consequent rough surface tending to produce continuous abrasion and heating. In bearings of this kind, the boxes or bearings should be of hard composition, as, say, a mixture of 12 parts of copper to $1\frac{1}{2}$ of tin, and $\frac{1}{2}$ of zinc, which will turn in the lathe easily, and yet be sufficiently hard to resist abrasion under ordinary duty.

ANNEALING OR SOFTENING.

To soften finished iron or steel work without damaging its finish, well lute an iron box with fire-clay, and place the work in it, surrounded by turnings or borings of the same metal as itself. Fill the box full of such turnings, place the lid on, and lute it with fire-clay; then place it in a furnace, heat slowly to a red heat, and allow the furnace fire to go out and the box to cool in the furnace.

To anneal electro-magnets, first heat the iron to a very low red heat, and let it cool off in soft soap; then reheat to a low red, and let it cool while well covered in slaked lime.

To anneal ordinary steel, heat it to a low red heat and allow it to cool in ashes or lime.

To remove the sand and scale from iron castings, immerse them in a pickle composed of one part oil of vitriol to three parts of water. In six to ten hours remove and wash them with clean water; when time is no object, make the solution weaker, and let them pickle longer.

MIXTURES OF METALS.

Babbitt metal bearings for fast running journals—Tin, 50 parts; antimony, 5 parts; copper, 1½ parts.

Brass for journal boxes—Copper, 10 pounds; tin, 1½ pounds; spelter, ½ pound.

Brass for valves—Copper, 9 pounds; tin, 1 pound; spelter, ¼ pound.

Bell metal—Copper, 50 pounds; tin, 11 pounds.

Yellow brass—Copper, 20; zinc, 10 pounds; lead, ¼ pound.

Yellow brass for castings—Copper, 36 parts; zinc, 17 parts; lead, 2½ parts; tin, 2½ parts. The zinc to be added last to prevent its burning away.

Gun metal—Copper, 9 parts; tin, 1 part.

Solders—Fine, tin 1 part, lead 1 part. Plumbers, tin 1 part, lead 2 parts. For cast-iron, tin 2 parts, lead 1 part.

Soldering liquid—Muriatic acid which has dissolved scraps of zinc until sponge zinc will form in it, is fit for soldering brass or copper work; for cast or wrought-iron work, sal-ammoniac should be added; while for tin, the latter may be omitted, and water, in the proportion of one-third, added.

CHAPTER XIII.

TAPS AND DIES.

TAPS should be forged of hammered square bar steel, and forged to as near the finished size as possible (so that they are large enough to true up), because the metal on the outer surface of a forging is, from receiving the most of the effects of the forging, of finer quality than the interior metal. After the forging of the tap is complete, it should be heated to a low red heat and covered in lime or ashes, the object being not only to soften the metal and make it easier to cut, but to release any tension there may be upon the outer skin, in consequence of the forging, for there is a tension upon the surface of all forged as well as cast work.

The effect of blows delivered upon forged work by the blacksmith's tools is not only greater upon the exterior than upon the interior of the metal, but is greatest upon that part of the forging which receives the most working, and upon that part which is at the lowest temperature during the finishing process: because the blows delivered during the finishing process are lighter than those during the earlier stages of the forging, and hence their effects do not penetrate so deeply into the body of the metal. Then again, on that part of the metal which is coolest, the effects of the light hammering do not penetrate so deeply; and from these combined causes, the tension is not equally distributed over the whole surface of the forging, and hence its removal, by cutting away the outer surface of any one part, and thus releasing the tension of that part, alters the form of the whole body, which does not, therefore, assume its normal shape until the outer skin of its whole surface has been removed. While the metal is at about an even heat all over, and is above a red heat, the effect of working the metal by forging it is simply, as already stated, to improve its texture, to close the grain, and thus to better its quality, especially toward and at its outer surface; but as the tension commences, while and after the metal loses its redness, we adopt the plan, after forging, to heat the tap all over to a low red heat, and to then lightly file its surface so as to remove any protruding scale; then allow it to cool of itself, without any forging being performed upon it at that heat. This process will nearly, if not entirely, remove the tension created by the forging.

If the tap is a long one, many experienced blacksmiths state that it should, after the forging is completed and the tap is very nearly or quite cold, be stood endwise on the anvil, and placing a flatter on the end of the tap, strike the flatter a sharp blow with a light sledge, which it is stated will set the metal so that it will not warp in the process of hardening. An excellent plan to effect the same object is to rough out the tap all over, so as to remove the tension, and to then heat the tap to a low red, and allow it to cool gradually.

The threads of taps of the smaller sizes should be finished by a chaser, so as to insure correctness in the angles and in the depth of the thread.

The taper tap should not be given more taper than the depth of the thread in the length of the tap, or it is liable to be used on holes that are too small, which places more duty upon it than is necessary and than it should be required to perform; rendering it, in consequence, liable to break from the excessive strain, and causing the square end of the tap, where the wrench fits, to twist and the corners to become rounded.

A tap which has much clearance placed upon its thread, by the screw-cutting tool, or by a chaser, will cut very freely, and will answer for rough work; but such a

Fig. 192.

tap does not cut a really good thread, and generally leaves the diameter of the thread in the hole larger than the diameter of the tap itself, because the tap is liable to wabble, and the least excess of pressure, on one end of the tap wrench more than on the other, causes the tap to lean towards the end of the wrench receiving the most pressure, and hence, to tap a hole larger than itself. Especially is this liable to occur if the tap wrench has more than one square hole in it so as to enable the same wrench to be used on more than one size of tap; for in such a case, the holes being not in the centre of the wrench, the weight of the wrench and the pressure placed on the end of the wrench will exert more pressure on one side of the tap than the other, in consequence of their greater distance or longer leverage from the tap. The same effects (from the use of such wrenches) are experienced in using taps having no clearance in the thread; but the thread in this latter case is so much nearer a fit to the hole that it serves as a guide and keeps the tap steady.

Then, again, if a tap has much clearance upon the thread, and is required to back out from and not pass entirely through the thread, the cuttings jam between the thread in the hole and the thread upon the tap, especially if the hole is in cast-iron. The sides of the teeth of such a tap have a very small bearing upon the sides of the thread in the hole, causing the tap, if used by hand, to work very unsteadily.

Taps for use in machines, such as the nut tap shown in Fig. 192, which are intended to pass through the work, may have considerable clearance in the thread, the greater part of which clearance should be at the back end of the teeth.

TAPS AND DIES.

Taps for ordinary work may have a very slight amount of clearance placed upon the teeth back from the cutting edge, just sufficient in fact to prevent them from binding *hard* against the sides of the thread being cut, and yet not sufficient to prevent the sides of the teeth from acting as a guide to steady the tap in the hole being threaded or tapped. Taps for holes requiring to be unusually exact in their diameter should not have any clearance placed upon the sides of the thread, but may have a flat place filed along the tops of each row of teeth, the flat face terminating close to the flute on either side of the length of the teeth, but in no case extending entirely to the flute. The flute of a tap should be volute and not carried over the back end of the teeth, otherwise the cuttings are apt to jam when the tap is being backed out.

If the flute of a tap is made spiral, it serves to steady the tap in the hole, especially if the latter is not round; but the extra trouble involved in making spiral flutes has prevented their universal application.

The taps shown in Fig. 193 represent a form of tap not

TAPER.

PLUG.

infrequently made, but which is wrong, because of having taper in the diameter of the bottom of the thread.

The taper of a hand or machine-tap, for all save gas

taps, should be turned parallel the same as a plug tap, and then have the taper made by turning off the thread to a straight taper which just turns the thread out at the entering end of the tap, and leaves about four full threads at the end of the tap thread; and such a tap will work much more steadily than one having more thread on the taper end.

Taps having thread on the small end of the taper tap do not cut a correct angle of thread at starting, and gradually right themselves as the tap enters, the reason for which will be found illustrated in Fig. 205, and in the remarks upon adjustable screw dies.

The plain part of a tap, that is, that part from the thread to the end of the square where the wrench fits, should be turned down a little smaller in diameter than the bottom of the thread (unless in the case of very small taps), so that the tap can pass right through the hole in all cases where the hole passes through the work, thus saving time by obviating the necessity of winding the tap back, and furthermore preserving the cutting edges of the tap teeth by avoiding the abrasion caused by their being rubbed backwards against the metal of the hole. For special work, where the holes to be tapped do not pass through the work, and it is therefore compulsory to wind the tap backwards to take it out of the hole, the plain part of the tap may be left larger than the diameter of the thread, the advantage being that the squares of several different sizes of taps may be made alike, and therefore to suit one tap wrench.

Taps for use in holes to be tapped deeply should be made slightly larger in diameter than those used to tap shallow ones, because in deep holes the tap is held steady by its depth in the hole, and because whatever variation there may be in the pitch of the threads in the hole and those on the bolt, is, of course, experienced to an extent greater as the length of the thread (that is, the number of threads) increases.

It is an excellent plan to finish the threads of a tap by

TAPS AND DIES.

passing it through a sizing die, that is, a solid die kept for that special purpose; but very little metal must be left on the tap for the solid die to take off, or it will soon wear and get larger. In making such a solid die, let its thickness be rather more than the diameter of the tap it is intended to cut, and make allowance for its shrinkage in hardening, for all holes shrink in hardening, while taps swell or become larger from that process; an allowance for this must therefore be made both in the case of the tap and the die. In the case of the solid die, it will be found that not only does the hole become smaller, but the external dimensions of the entire die have become larger by reason of the hardening, so that while the term shrinkage is correct, as applied to the hole, it is incorrect as applied to the die, the fact being that the metal of the die (the same as the metal of the tap) has expanded, extending its dimensions in all directions, and therefore in the direction of the centre of the hole, hence causing a decrease in its diameter or bore.

Three flutes are all that are necessary to small taps (that is, those up to about half an inch in diameter), which leave the tap stronger and less liable to wabble, especially in holes that are not round, than if it had four flutes. Taps of a larger size may have more flutes, but the number should always be an odd one, so that the tap will do its work steadily.

The United States standard for threads, which was first adopted by the Franklin Institute, is as follows:

DIAMETER OF TAP.

$\frac{1}{4}$	$\frac{5}{16}$	$\frac{3}{8}$	$\frac{7}{16}$	$\frac{1}{2}$	$\frac{9}{16}$	$\frac{5}{8}$	$\frac{3}{4}$	$\frac{7}{8}$	1	$1\frac{1}{8}$	$1\frac{1}{4}$	$1\frac{3}{8}$	$1\frac{1}{2}$	$1\frac{5}{8}$	$1\frac{3}{4}$	$1\frac{7}{8}$	2

NUMBER OF THREADS TO INCH.

20	18	16	14	13	12	11	10	9	8	7	7	6	6	$5\frac{1}{2}$	5	5	$4\frac{1}{2}$

In this standard, the screw threads are formed with straight sides at an angle of sixty degrees to each other, having a flat surface at the top and bottom equal to one-eighth of the pitch, the pitches as above.

The English or Whitworth standard varies from the above both in shape and number of threads to inch, as below:

DIAMETER OF TAP.

| $\frac{1}{4}$ | $\frac{5}{16}$ | $\frac{3}{8}$ | $\frac{7}{16}$ | $\frac{1}{2}$ | $\frac{5}{8}$ | $\frac{3}{4}$ | $\frac{7}{8}$ | 1 | $1\frac{1}{8}$ | $1\frac{1}{4}$ | $1\frac{3}{8}$ | $1\frac{1}{2}$ | $1\frac{5}{8}$ | $1\frac{3}{4}$ | $1\frac{7}{8}$ | 2 |

NUMBER OF THREADS TO INCH.

| 20 | 18 | 16 | 14 | 12 | 11 | 10 | 9 | 8 | 7 | 7 | 6 | 6 | 5 | 5 | $4\frac{1}{2}$ | $4\frac{1}{2}$ |

In this standard, the screw threads are formed with flat sides at an angle of fifty-five degrees to each other, with a rounded top and bottom. The proportions for the rounded top and bottom are obtained by dividing the depth of a sharp thread having sides at an angle of fifty-five degrees, into six equal parts, and within the lines formed by the sides of the thread, and the top and bottom dividing lines inscribing a circle, which determines the form of top and bottom of thread.

Taps should be heated, for hardening, in a charcoal fire, and be heated slowly to a cherry red, and then dipped perpendicularly into clean water. The water should be made sufficiently warm to feel pleasant to the hand; for, if the water has not the cold chill taken off it, the taps are apt to crack along the flutes. The tap should be lowered perpendicularly in the water, even after it has disappeared below the surface; but in no case should it be moved sideways, or it will warp. It should not be taken out of the water until quite cold, or it will crack after it is taken from the water and during the cooling process. After the tap is hardened, it should be brightened along the flutes and on the plain part, and then lowered, as follows: A piece of tube, about half the length of the tap, and of about twice or three times its diameter, and having its thickness about the same, if possible, as the diameter of the tap, should be heated in the fire to an even cherry-red heat, and then taken from the fire and placed in such a position that it is open to clear daylight and not affected by the rays of light from the fire.

The tap should be held in a pair of tongs, whose jaws have been well warmed; and a small piece of metal should be interposed between the jaws of the tongs and the sides of the square of the wrench-end of the tap, so that the tongs may not obstruct the square of the tap from receiving the heat from the tube. The tap and tongs should then be passed through the heated tube, so that the square end of the tap and the tongs only will be inside the tube. The tap should be slowly revolved while in this position; and when the tap has at that end become slightly heated, but not enough to draw the color, the shank and threaded part of the tap should then be slowly passed endways back and forth, and, while slowly revolving through and through the centre of the tube, until the color appears, and if it appears of an even hue all over, proceed until a brown color appears; then withdraw the tap from the tube and quench it perpendicularly in warm water. If, however, the color does not not appear so quickly in any particular part, hold that part in the tube a little the longest, and if either end lowers too rapidly, cool it by a slight application of oil. The square end of the tap, on which the wrench fits, may be lowered to a deeper color, as may also the shank of the tap, than the threaded part, which will leave them stronger and less liable to twist or break. By using the size of tube here recommended, it will be found that the tempering process will be performed, and the colors appear very slowly, so that there will be ample time to judge when the precise requisite degree of hardness has been reached. This plan is far superior to tempering in heated sand. Very long taps may be greased and heated preparatory to being hardened in molten lead, the object being to heat the outside of the tap evenly all over to a red heat so rapidly that the inside

246   COMPLETE PRACTICAL MACHINIST.

metal of the tap is comparatively cool, hence, when the tap is hardened, the outside only is hardened; and, if the tap warps in the hardening, it can, after being tempered, be

Fig. 194.

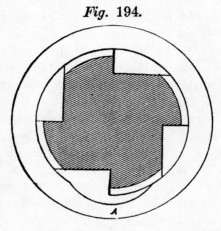

straightened, the soft metal of the centre of tap preventing it from breaking in the straightening, which should be per-

Fig. 195.

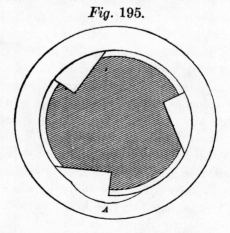

formed with a leaden hammer, and with the tap resting upon lead.

In the Whitworth or English standard, taps are given

three flutes, which has the merit that the thread will be cut more nearly axially true with the hole, notwithstanding that the hole may be out of round, or have a blow hole on one side.

Thus in Fig. 195 we have a three-flute tap in a hole out of round at A, and it is obvious that when a cutting edge meets the recess at A, all three teeth will cease to cut; hence there will be no inducement for the tap to move over toward A. But in the case of the four-flute tap in Fig. 194, when the teeth come to A there will be a strain tending to force the teeth over toward the depres-

Fig. 196.

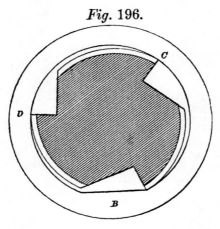

sion A. How much a given tap would actually move over would, of course, depend upon the amount of clearance; but whether the tap has clearance or not, the three-flute tap will not move over, while with four flutes the tap would certainly do so. Again, with an equal width of flute there is more of the circumference tending to guide and steady the three-flute and the four-flute tap. If the hole has a projection instead of a depression, as at B, Figs. 196 and 197, then the advantage still remains with the three-flute tap, because in the case of the three flutes, any lateral movement of the tap will be resisted at the two points C and D, neither of which are directly opposite to the location of the projection B; hence, if the projection caused the tap to move laterally, say, $\frac{1}{100}$ inch, the effect at C and D would be very small, whereas in the four-flute, Fig. 197, the effect at E would be equal to the full amount of lateral motion of the tap.

It is quite true that the four teeth will cut easier than the three, but against this we have the fact that the three will tap a more parallel hole because it will work steadier. The length of the teeth and the width of flute would obviously influence the steadiness of the tap, especially if its teeth

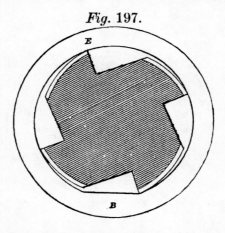

Fig. 197.

are a true circle, and therefore have no clearance. In Whitworth's taps the flutes are narrower, and the teeth longer than in American practice, producing more parallel and round holes, but requiring more power to operate them. Fig. 198 represents the Whitworth form, Fig. 199 the Brown & Sharpe, and Fig. 200 the Pratt & Whitney for hand taps; average American practice being represented in Fig. 201. The Brown & Sharpe, it will be observed, has the shortest teeth and the most shallow flute, and is the strongest.

In hand taps the position of the square with relation to

TAPS AND DIES.

the cutting-edges is of consequence; thus, in Fig. 202. there being a cutting-edge, A, opposite to the handle, any

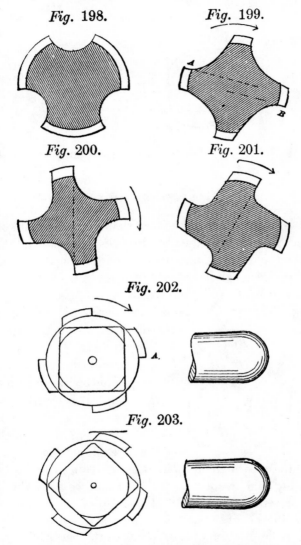

Fig. 198.

Fig. 199.

Fig. 200.

Fig. 201.

Fig. 202.

Fig. 203.

undue pressure on that end of the handle would cause A to cut too freely and the tap to enlarge the hole; whereas in Fig. 203 this tendency would be greatly removed, and still more removed in Fig. 204, the disadvantage of

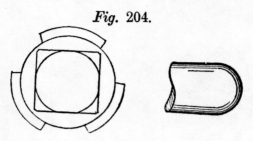

Fig. 204.

that in Fig. 202 being greatest when a single-ended wrench is used.

ADJUSTABLE DIES,

That is, those which take more than one cut to make a full thread, should never be used in cases where a solid die will answer the purpose, because adjustable dies take every cut at a different angle to the centre line of the bolt, as explained by Figs. 205 and 206.

Fig. 205 represents an ordinary screw. It is evident that

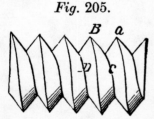

Fig. 205.

the pitch from a to B is the same as from C to D. Let $a\ b$, in Fig. 206, represent the centre-line of the bolt lengthwise, and $c\ d$ a line at right angles to it: then let from the point e to the point f represent the circumference of the top of the thread, and from e to g the circumference of the bottom of the thread, the lines $h\ h$ representing their respective pitches; and we have the line k, as representing the angle of the top of the thread to the centre line $a\ b$, of the bolt, and the line l, as representing the angle of the bottom of the thread to the centre-line $a\ b$, of the bolt, from which it becomes apparent that the top and the bottom of the thread are at different angles to the centre line of the bolt.

TAPS AND DIES.

The tops of the teeth of adjustable dies are themselves at the greatest angle, while they commence to cut the thread on the bolt at its largest diameter, where it possesses the

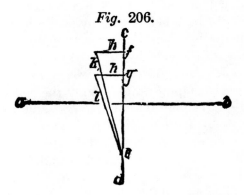

Fig. 206.

least angle, so that the dies cut a wrong angle at first, and gradually approach the correct angle as they cut the depth of the thread.

From what has been already said, it will be perceived that the angle of thread, cut by the first cuts taken by adjustable dies, is neither that of the teeth of the dies nor that required by the bolt, so that the dies cannot cut clean because the teeth do not fit the grooves they cut, and drag in consequence.

DIES FOR USE IN HAND STOCKS

are cut from hubs of a larger diameter than the size of bolt the dies are intended to cut; this being done to cause the dies to cut at the cutting edges of the teeth which are at or near the centre of each die, so that the threads on each side of each die act as guides to steady the dies, and prevent them from wabbling as they otherwise would do; the result of this is, that the angle in the thread in the dies is not the correct angle for the thread of the bolt, even when the dies are the closest together, and hence taking the finishing cuts on the thread, although the dies are nearer the correct angle when in that position than in any other. A very little practice at cutting threads with stocks and dies will demonstrate that the tops of the threads on a bolt, cut by them, are larger than was the diameter of the bolt, before the thread was commenced to be cut, which arises from the pressure, placed on the sides of the thread of the bolt, by the sides of the thread on the dies, in consequence of the difference in their angles; which pressure compresses the sides of the bolt thread (the metal being softer than that of the dies) and causes a corresponding increase in its diameter. It is in consequence of the variation of angle in adjustable dies that a square thread cannot be cut by them, and that they do not cut a good V thread.

In the case of a solid die, the teeth or threads are cut by a hub the correct size, and they therefore stand at the proper angle; furthermore, each diameter in the depth of the teeth of the die cuts the corresponding diameter on the bolt, so that there is no strain upon the sides of the thread save that due to the force necessary to cut the metal of the bolt thread.

In a taper tap, whether it have a V thread or a square one, each individual diameter and angle of thread cuts the same diameter and angle in the hole, providing the bottom of the thread is of the same diameter all along the tap, and that the taper is made by turning off the tops of the threads; if, however, the tap is made taper and the thread is cut taper, the angles of the sides of the thread itself will not stand true with the diameter of the body of the bolt or tap, nor will the angle of the thread stand true with the centre line of the length of the bolt.

A solid die, having the teeth tapered off so that at the entrance there will be the bottom of the thread only and a full thread will not have been reached until the bolt has entered the die to a distance equal to five or six times the amount of the pitch of the thread, will cut a moderately good square thread, but such dies take a great deal of power to drive them.

In making dies, whether they are adjustable or not, it is of the utmost importance that the recesses, or spaces cut to form the cutting edge of the teeth, be roomy, so that the cuttings will pass easily away and not clog in the die, as is too commonly the case. The teeth of a die may be given a little top rake.

Fig. 207 represents a pair of stocks and dies of the pattern in ordinary use, there being to each pair of stock dies from ⅛ to ¼, inclusive, by 32ds. Similar stocks

Fig. 207.

with keys instead of pins, and containing dies up to 1⅛ inches, are supplied by the same firm.

In the Whitworth stocks and dies there are two cutting dies or chasers and one guide die, the construction being as follows.

In Fig. 208, A is the guide die, and B C are the chasers, which are set in to the cut by a key having bevels at a P. The two chasers B and C are moved by the key in lines that would meet at D, and, therefore, at a point that is behind the centre of the work, and this has the effect of preserving their clearance. It is obvious, for example, that as these chasers cut a thread on the work the latter will move over in the dies toward chaser A on account of the thread on the work sinking into the threads on A, and this motion would prevent the chasers B C from cutting if they moved in a line pointing to the centre of the work. This is more clearly shown in Fig.

209, in which the guide die A and one of the cutting dies or chasers B is shown removed from the stock, while the

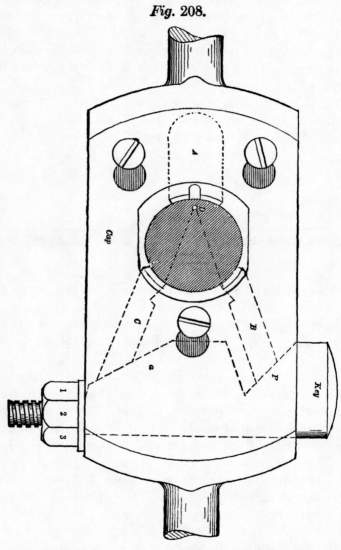

Fig. 208.

bolt to be threaded is shown in two positions—one when the chaser B is at the circumference of the work, and the other when it has entered to the full thread depth, the

Fig. 209.

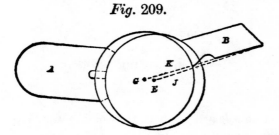

points or dots. E and G, representing the respective centres of the work tor the two positions.

CHAPTER XIV.

VISE-WORK—TOOLS.

CHISELS.

FLAT chisels are made from two shapes of bar steel, one of which is shown in Fig. 210, and the other in Fig. 210 a.

Fig. 210. Fig. 210 a.

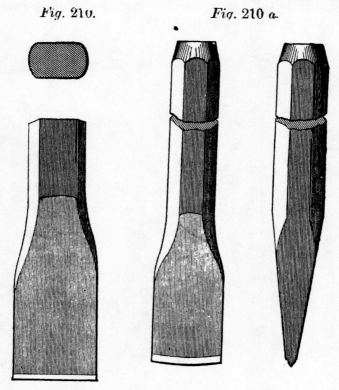

The difference between the two is, that as the cutting edge wants to be parallel to the flats on the chisel, and as Fig. 210 has the widest flat, it is easier to tell when the cutting edge and the flat are parallel, and the broad flat is the best guide in holding the chisel level with the surface to be chipped. Either of these chisels is of a proper width for wrought-iron or steel, because chisels used on these metals take all the power to drive that can be given with a hammer of the usual proportions for heavy chipping, which is: weight of hammer 1¾ lbs., length of hammer handle 15 inches, the handle to be held at its end and swung back about vertically over the shoulder. If we use so narrow a chisel on cast-iron or brass and give full-force hammer-blows, it will break out the metal instead of cutting it, and the break may come below the proper depth and leave ugly cavities; so for these metals the chisel may be made broader, as in Fig. 211, so that the force of the blow will be spread over a greater length of chisel edge and will not move forward so much at each blow, and therefore it will not break the metal out. Another ad-

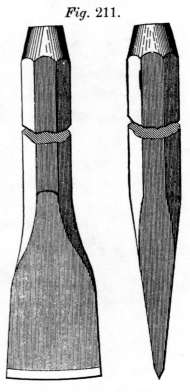

Fig. 211.

vantage is, that the broader the chisel the easier it is to hold its edge fair with the work surface, and make smooth chipping. The chisel point must be made thin as possible, the proportions shown in the figures being suitable for new chisels. In grinding the two facets to form the chisel we must be careful to avoid grinding them rounded, as shown in A in the magnified chisel ends in Fig. 213, the proper way being to grind them flat as at B in the sketch.

Fig. 212.

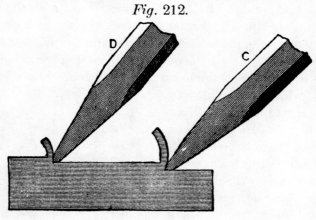

Fig. 213.

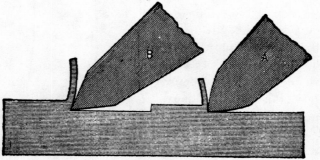

Fig. 214.

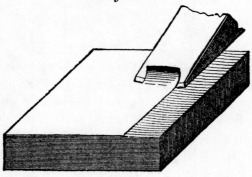

We must make the angle of these two facets as acute as

possible, because the chisel will then cut easier. The angle at C in Fig. 212 is about right for brass, and that at D is about right for steel. The difference is that with hard metal the more acute angle dulls too quick. Considering now the length of the cutting edge, it may for heavy chipping be made straight, as in Fig. 210, or curved, as in Fig. 210a, which is the best because the corners are relieved of duty and are therefore less liable to break. The advantage of the curve is greatest in fine chipping because, as is seen in Fig. 214, a thin chip can be taken without the corners cutting, and these corners, being exposed to view, aid the eye in keeping the chisel edge level with the work surface. In any case

Fig. 215.

Fig. 216.

Fig. 217.

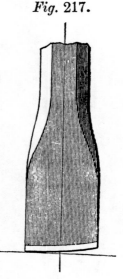

we must not grind it hollow in its length, as shown in Fig. 215, and shown exaggerated in Fig. 216, because in that case the corners will dig in and cause the chisel to be beyond our control, and, besides that, there will be a force that, acting on the wedge principle and in the direction of the arrow, will act to spread the corners and break them off.

260 COMPLETE PRACTICAL MACHINIST.

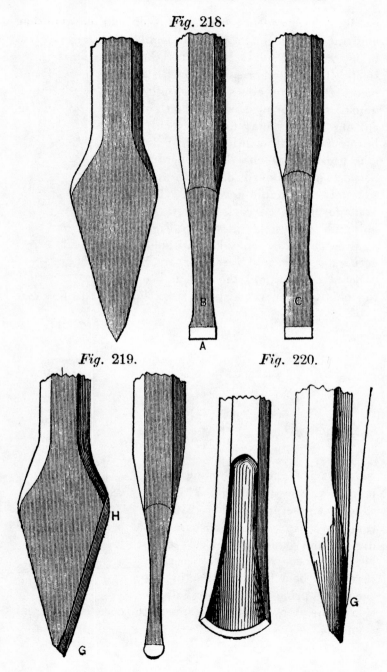

Fig. 218.

Fig. 219.   Fig. 220.

The facets must not be ground wider on one side than on the other of the chisel, as in Fig. 215, because in that case the flat of the chisel will not form a guide as to when the cutting edge is fair and level with the work surface. The length of the cutting edge should be at a right angle to the chisel body, for if ground to one side, as in Fig. 217, the chisel will be apt to jump sideways at each blow.

Fig. 221.

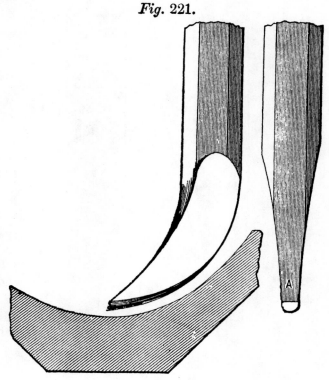

When a heavy chip is to be taken over a surface broader than the chisel, the cross cut cape chisel, shown in Fig. 218, is brought into requisition, being employed to cut grooves, as shown at A B and C in Fig. 224, these grooves being spaced apart a little less than the chisel width, so as to relieve the chisel corners from duty, and thus prevent them from breaking. The point A should be tapered back to B, so that the chisel may be moved laterally to keep the cut in a straight line, which could not be done if the end were made parallel, as at C in the figure.

The round-nosed chisel, Fig. 219, may be made straight

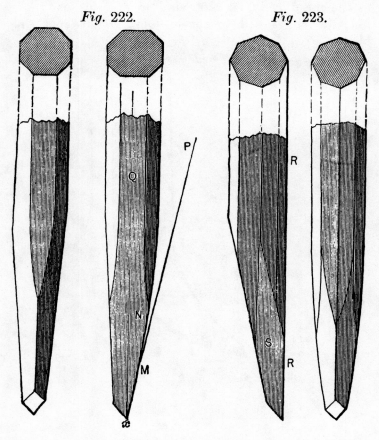

Fig. 222.   Fig. 223.

from H to G, but should be bevelled from G to the point, so that the depth of the cut may be altered at will by raising or lowering the chisel head, and the same remarks apply to the cow mouth chisel, in Fig. 220. The oil groove chisel shown in Fig. 221 is employed to cut oil grooves in the bores of brasses or journal boxes, and should have a less curvature than the bore of the brass or box. It should be made thinner at A than at the cutting edge, to prevent any wedging action, which would raise a burr on each side of the oil groove.

Fig. 224.

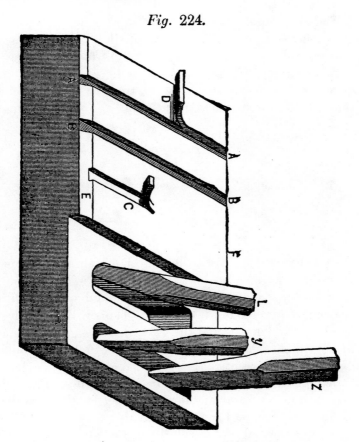

The diamond point chisel, shown in Figs. 222 and 223, is for cutting out square corners, and is varied in shape to suit the work. When the hole in the work is not deep, the chisel can be bevelled at M, and thus bring its point X central to the body of the steel, as shown by the dotted line Q, rendering the corner X less liable to break, which is the great trouble with this chisel. But as the bevel at M necessitates the chisel being leaned over, as at Y in Fig. 224, it could not be kept to its cut in deep holes; so we must omit the bevel at M, and make that edge straight as at R R in Fig. 223.

The side chisel obeys just the same rule, so we must give it bevel at W in Fig. 225 for shallow holes and lean it over, and leave it very nearly straight for deep ones.

Applications of chisels are shown in Fig. 224, the cut of a flat one being shown at D, the cow mouth being shown at L, the diamond point at Y, and the side chisel at Z.

Fig. 225.

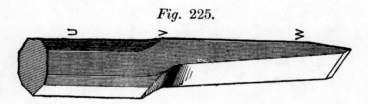

In tempering a chisel, the end should first be dipped about half an inch in the water and held still; it should then be dipped another half an inch and quickly withdrawn and lowered down to a blue. This will give a deep band of blue temper, and the chisel may be ground many times without losing its temper, and it will be less liable to break. In exceptional cases the temper may be left a little higher, as say a reddish purple for hard steel; but in this case, heavy hammer blows must not be given, as they will break the chisel. To prevent the work from breaking out at the end of the chisel cut, the edges of the work should be bevelled or chamfered off as in Fig. 224 at E and F.

CALIPERS.

Outside calipers, that is, those used for measuring external diameters, should have larger rivets in them than they

VISE-WORK—TOOLS.

are generally given, a fair proportion being a rivet of one-half inch diameter for a pair of calipers intended to measure up to diameters of seven inches. The points of such calipers should be tapered to a wedge shape, the tapering face being on the outside edge, so that the same part of the points of each leg will touch the work, whether the latter is of small or large diameter; the points, where they meet together, should be slightly rounding, so that they will touch the work at the middle of each point.

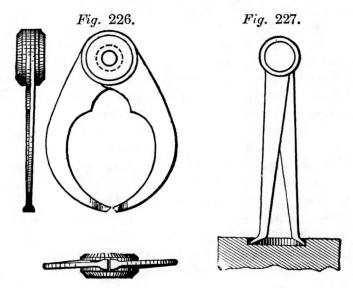

Fig. 226. Fig. 227.

Figs. 226 and 227 represent excellent shapes for inside and outside calipers.

The short angular ends of the inside calipers enable them to enter a long distance in small holes.

The spring calipers sold at the stores serve very well for rough measurements, but are not suitable for very fine ones. Furthermore, when operated by a wing nut, the same is apt to be moved in laying the calipers down. The forms, however, in which the legs are moved by a right and left hand screw are excellent when means are provided for locking the screw.

Another feature to the advantage of this form is, that when the legs are extended, the points are still at the extreme end of the calipers, so that the points will measure to the extreme end of the hole, even though the latter is closed by metal, that is, terminates in the metal. This is not the case when the caliper ends are bent round to the usual extent, for the curve of the bend will touch the end of the hole and prevent the caliper points from reaching it. In measuring with calipers, let the points be set to touch the work very lightly indeed, or they will spring from the pressure due to forcing them over the work.

Compass calipers should have the end of one leg bent round at the end, about the same as is customary for inside calipers, the other leg being pointed like a compass leg. They are valuable tools, and may be employed to mark off the centres of holes, or to try if a centre already existing is in the exact centre of the hole. Or they will mark off a face, so that it will fit another face, whether it be regular or irregular, the curved point being kept against the irregular face, and the point describing (by moving the compass along) a similar line on the face to be fitted. They will answer for many of the uses to which a scribing block is put; and being lighter and more easily handled, and, furthermore, capable of doing duty without the use of a surface or scribing plate, they are in such cases far preferable.

The legs may be crossed so that the curved point inclines to the straight point, in which position they will mark the centres of shafts or rods, either round, square, or any other shape, or try such centres, when they already exist, more accurately than can be done by any other tool. They will, in this case, mark off a line at the distance to which they are set round any surface; they are employed to mark off keyways, or the taper of a gib when the key and one edge of the gib is placed, and for a variety of other uses too numerous to recapitulate, being among the most useful tools the fitter can possibly possess. The points of calipers should be tempered to a blue, and of compass calipers to a straw color.

THE SQUARE.

The square is too common a tool to require any description of its form. The best method to make one is to make the back of steel, and in two halves, one-half being the thickness of the blade thicker than the other. The slot for the blade must then be filed in the thickest half, to the depth exactly equal to the thickness of the blade. The two halves composing the back must then be riveted together, and the edges surfaced each true of itself (using a surface plate to try them), and also true with each other. The blade, which should be made of saw blade, may then be put into its place, ready to have the holes for the rivets drilled. It should be placed so that the outer end is a little depressed (on the inside angle) from the right angle; this is done so that whatever there may be to take off the blade (after it is riveted to the back), to make its edges form right angles to the back, will require to be taken off the outer end of the inside angle and the end of the blade forming the corner of the outside angle, so that no work will require performing on the blade in the corner formed by the blade entering the back on the inside angle, where it would be difficult to file or scrape without injuring the edge surface of the back. The best way to true a square is to turn up a piece of round iron equal in length to the square blade, being careful to make it quite parallel, and then true up the end of the iron, making it hollow towards the centre, or cutting it away from the centre to within an eighth of an inch of its diameter, so that it will stand steadily on its end. If the piece of iron be then stood on its end on a surface plate, its outline on each side, which represents its diameter, will form a true right angle to the surface of the plate, and hence a gauge with which to true the square.

THE SCRIBING BLOCK.

This tool is made in a variety of forms, but the simplest and best form is that shown in Fig. 228, in which D is the block complete, the scriber being a simple piece of round steel wire. The dotted line on the foot is the distance to which the foot is hollowed out to make it stand firm. Fig. E is the bolt and nut; the bolt has a flat side filed on each side of it to fit it to the slot in the scribing block stem, so that the bolt cannot turn when

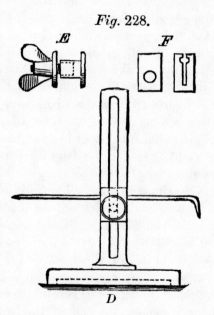

Fig. 228.

it is being tightened. Fig. F is a face and edge view of the piece or clamp for the scriber which passes through the hole in the slot.

The advantages possessed by this form over other forms of scribing block are that it is easy to make, and that the scriber, being a piece of wire, is easily renewed. It holds the scriber very firmly indeed, and the scriber may be moved back and forth without the nut becoming slack.

FILES AND FILING.

Large files should be fitted to their handles by making the tine of the file a low red heat, and forcing it into the handle so that it will burn its way into the handle, and thus prevent the handle from splitting, as it would do if the file tine were driven in; the file and handle should be turned in the hands occasionally to guide the eye in detecting whether the file is entering in a line with the length of the handle. Care should be taken to wrap a piece of waste around the end of the file, and to keep it wetted with water so as to avoid softening the teeth of the file while heating the tine. For small files it is sufficient to bore a small hole in the handle and force the tine in by hand. A file should be held so that the butt end of the file handle presses against the centre of the palm of the hand, the forefinger being beneath the body of the file handle.

In selecting a file, choose one that is thickest in the centre of its length, and of an evenly curved sweep from end to end, so as not to make the surface of the work round by filing away the edges. Files that have warped in the hardening may be used on very narrow surfaces, or on round or oval work; or, if they are smooth files, they may be used on lathe work. Keyways or slots, especially, require an evenly rounded file; and if the keyway is long and the file parallel or uneven upon its surface, the end of the file only should be used to ease away the centre of the keyway or the high spots. It is also highly advantageous to rub chalk on the teeth of the file, so that, after a little using, the eye can detect the part of the file which is highest, and govern its use accordingly.

Half-round files should be rounded lengthwise of the half-round side of the file, because it is difficult to file out a sweep evenly, even with a well-shaped file, and it is impossible to do so with a file whose half-round surface is hollow in the direction of its length.

Files should be held as in Fig. 229, the end of the handle abutting against the palm of the hand, and on broad surfaces should be used with a lateral sweep at

Fig. 229.   Fig. 230.

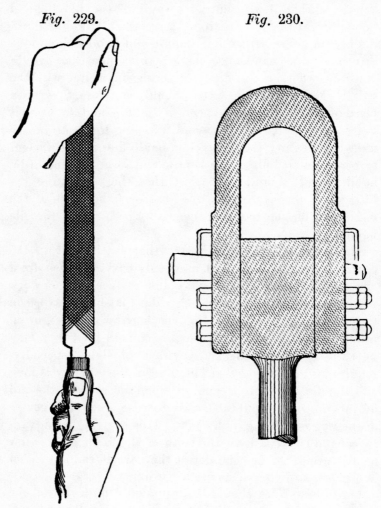

each stroke, so that the file marks will cross and recross, as in Fig. 230. When a raised handle, such as in Fig. 231, is used, the most rounded or bellied side of the file should be applied to the work, so as to confine the duty to as few teeth as possible, and thus enable the file to cut more freely.

Fig. 231.

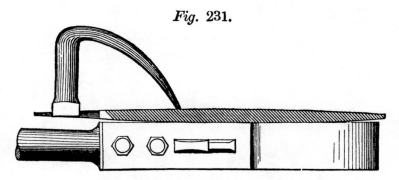

In using a half-round file, to file out a curve, the file should be given a lateral sweep, first from right to left, and then from left to right, changing the direction after

Fig. 232.

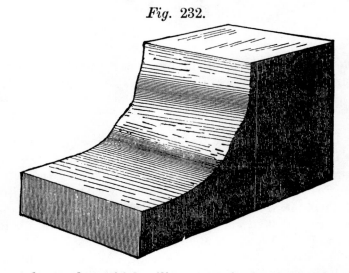

every few strokes, which will prevent the formation of such waves as are shown in Fig. 232. And in draw filing, the unit should so bend at each file stroke that different parts

of the file meet the work at every part of its motion, as is
shown in Fig. 233. At the same time, however, it is
necessary, at each stroke, to move the file endways a trifle
front in one direction and then in the other, so that the

Fig. 233.

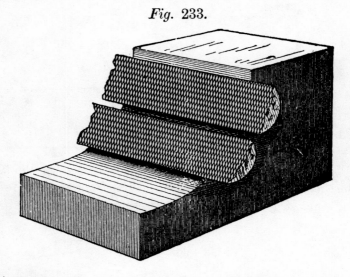

file marks cross each other, and thereby produce an even
curvature on the work.

FILING OUT TEMPLATES.

Let it be required to make or test a piece of work such
as in Fig. 234, the teeth to be equally spaced, of the same
angle, and of equal height. A template must be made of
the form shown in Fig. 235. To begin with, take a piece
of sheet metal equal in width to at least two teeth, and
assuming that the template is to have two teeth, file its
sides, P Q, in Fig. 236, parallel, and make the width
equal to twice the pitch of the teeth. We next divide its
width into four equal parts by lines, and mark the height,
as shown in Fig. 236. If we desire to make the template
such as at 1 in Fig. 235, we cut out the shaded portion

at A, Fig. 236; or for the template at 2, Fig. 235, the shaded portion at B. It will be observed, however, that in template A, Fig. 236, there are two corners, C and D, to be filed out, while at B there is but one (E), the latter

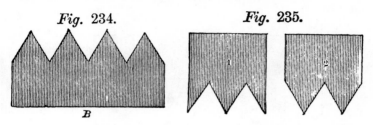

Fig. 234. Fig. 235.

being the easier to make, since the corners are the most difficult to file and keep true. The best method of producing such a corner is to grind the teeth off with a half-round file on the rounded side, producing a sharper corner than the teeth possess, while giving at the same time a safe edge that will not cut one angle while the other is being filed. But when we come to apply these templates to the work, we shall find that A is the better of the two, because we can apply the square S (Fig. 237) to the outside of the template, and also to the edge F of the work, which cannot be done to the edges G of the work and H of the template, because the template edge overhangs. We can, however, apply a square, S', to the other edge of B, but this is not so convenient unless the tops of the teeth are level.

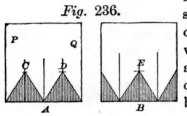

Fig. 236.

Assuming, therefore, that the template A is the one to be made, we proceed to test its accuracy, bearing in mind that for this purpose the same method is to be employed, whatever shape the template may be. Consequently, we make from the male template A, Fig. 238, a female template, K, Fig. 239, beginning at one end of K and filing it to fit A until the edges of A and K are in line when tested by a straight-edge, S. We then move the template

Fig. 237.

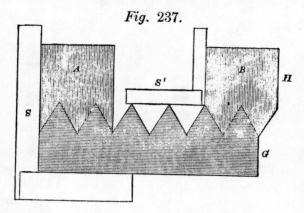

A one tooth to the right, and file another tooth in K, and proceed in this way until a number of teeth have been made, applying a square as at S, Fig. 238, to see that the

Fig. 238.

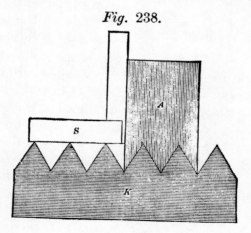

template A is kept upright upon K. When K has been thus provided with several teeth that would fit A in any position in which the latter may be placed, we must turn template A around upon K to test the equality of the angles. Thus, suppose at the first filing the edges *a b c d* upon A accurately fit the template K, and the straight-edge shows the edges fair. Then if we simply turn the template A around, its angles, which were before on the right, will now be on the left, as is shown at the right of Fig. 239. Thus in one position *a* fits to *e*, in the other it fits to *h*, or *b* fits to *f*, and when turned around it fits to *g*, and so on. Supposing that when thus turned around the

Fig. 239.

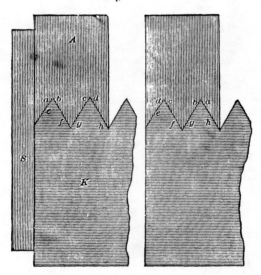

angles do not coincide, then half the error will be in the teeth of A, and one-half in those of K, and the best plan will be to correct them on A to the necessary amount as near as judgment will dictate, and then to apply K as before, continuing this process until A will fit anywhere in K, and may be turned around without showing any error. But at each correction the straight-edge must be applied, and finally should be tried to prove if the teeth tops are level. We have two interchangeable templates, of which

A may be used on the work and K kept to correct A when the latter becomes worn. It may be as well to add, however, that in first applying A to K it is best to press the straight-edge S against the edge of K, and hold it

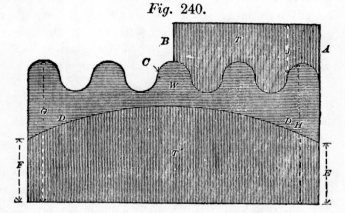

Fig. 240.

there, and then to place it against S, and slide it down into K.

Fig. 240 represents an example in which, the form being a curve, it would be best to have the template touch

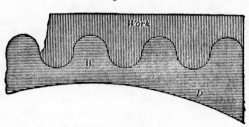

Fig. 241.

more than two teeth, as shown in the cut. By letting the side A of the template T terminate at the centre line of the two curves, and the end B terminate at the top of a curve, turning the template around would cause end A to envelope side C of the middle curve, thus increasing the scope of the template. Suppose, however, that the base curve D required to be true with the teeth then a second template T' must be used, its ends at E and F measuring an equal length or height, so that when they are placed even with the ends of the work, the distances G H being equal, the corrugations will be true to the curve D D.

Fig. 242.

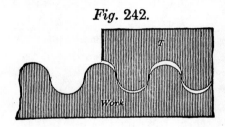

Now let it be supposed that, instead of making a template to test a piece of work such as W in Fig. 240, it is required to make a template for use in making another piece of work that is to fit to piece W; then template T in Fig. 240 will not answer, because it is a female template, whereas a male one is required, so that the edge

Fig. 243.

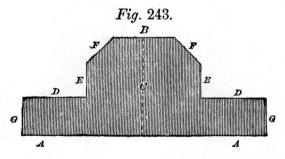

of the template may coincide with that of the work. But we may convert T (Fig. 240) into a male template by simply cutting off the edge A as far as the line J, leaving it as at T in Fig. 242, and causing its right-hand edge to coincide with the edge of the work so that the latter, after being fitted to the template, may be turned upside down and fit upon the piece of work W, Fig. 241, which corresponds to the piece W, Fig. 240. Fig. 243 represents an example in which the forms of both sides of a piece require to be exactly alike, and the easiest method of accomplishing this is as follows: The edge A should first be made true, and edge B made parallel to A. A centre line, C, may then be drawn, and from it the lines E E may be marked. The lines D are then drawn parallel to A A, lines E being made square to D and to A. The sides E may be calipered to width and parallelism, and all that will then remain is to file the angles F F and the ends G G to their required lengths. For F F all that is necessary is a template formed as in Fig. 244. The object of dressing the ends G G last is that if they were finished before, their faces E would have to be made at exactly correct distances from them, which would render the job considerably more difficult.

Fig. 244.

Fig. 245.

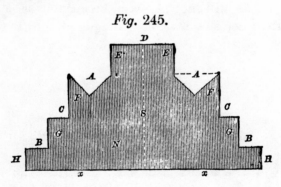

Fig. 245 represents a sketch for a piece of work whose two sides are to be shaped exactly alike, requiring a template of the form shown. From this a second template, M, Fig. 246, is made, and to this latter the work may be filed. To make the template in Fig. 245 the edge $x\ x$ must be made straight, and the edge D parallel to it at the proper height. A centre line, S, is then

VISE-WORK—FILING TEMPLATES.

marked, and the edges at E may be filed equidistant from S and square to D; hence they will be parallel to each other. The side sections F should then be filed equidistant from S and parallel to each other. They should be the proper width apart and square to D, being tested in

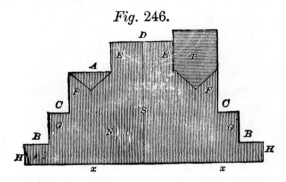

Fig. 246.

each of these respects. The line joining E and F should be left full, as denoted by the dotted line at A on the right. The edges at C C should then be filed, calipering them from the edge x x. Edges G G are obviously equidistant from S and parallel to S, or, what is the same thing, at a right angle to x x, from which they may therefore be tested with a square, and, finally, the edges B are made parallel to x x, and the ends H made square to x and equidistant from S. We have now to file the angular groove at A, and to get this true after marking its depth from the lines at A, we file it first to the lines as near as may be by the eye and very nearly to the full depth. We then make a small supplemental male template equal in width to the distance E F, or, in other words, to the width of the step at A, and having its edges quite parallel. Its end is then filed to fit the groove at A. when its edge meets and coincides with edge E, as

Fig. 247.

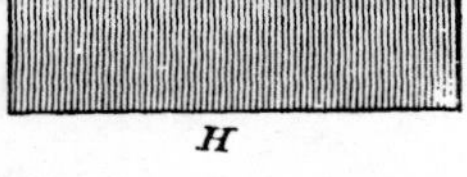

in Fig. 246, T representing the supplemental template. It is clear that when the V-groove A is so filed that T will fit it with either of its edges against E, the angles of the groove will be alike, and we may then make a male gauge, as in Fig. 247, that may be used to mark or line out the work and to use a template to file it to, its edge H being kept parallel to face D (Fig. 245) of the work.

SCRAPERS AND SCRAPING.

Surfaces requiring to be made very true may be finished by the hand-scraping process, but as filing is much quicker than scraping, the work should be filed as true as possible before the scraper is applied to it. Fig. 248 represents an

Fig. 248.

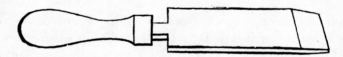

ordinary form of hand scraper, and Fig. 249 a form that will cut somewhat smoother. For use on wrought-iron, the cutting edge should be kept moistened, or it will tear the metal instead of cutting it cleanly. All surfaces intended to be scraped should first be filed as true as

Fig. 249.

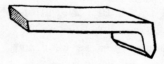

possible with a smooth file, care being exercised to use a file that is evenly curved in its length and slightly rounding in its breadth. After the surface has been scraped once or twice, a well-worn, dead smooth file may be passed over it, which will rub down the high spots

VISE-WORK—FILES.

of the scraper marks and greatly assist the operation of scraping. Scraping should be executed in small squares, the marks of one square being at a right angle to the marks of the next; then, after the surface plate has been applied, repeat the operation of scraping in squares, but let the marks cross those of the previous scraping. The face of the surface must be wiped off very clean before the surface plate is applied, or the surfaces of both the plate and the work will become scratched. The face of the plate may be moistened by the application of a barely perceptible coat of Venetian red, mixed with lubricating oil, rubbed on by the palm of the hand, to operate as marking to denote the high spots. In applying the surface plate, move it both ways on the work, and reverse it endwise occasionally. If the work is light, it may be taken from the vise and laid upon the plate; but much pressure need not be placed upon the work, or it will spring to suit the surface of the plate, and thus appear to be true when it is not so. Small surfaces should be rubbed on the outer parts of the surface of the plate, by which means the wear on the surface plate will be kept more equal.

VISE CLAMPS.

To prevent the jaws of a vise from damaging finished work, we require a pair, each, of lead and copper clamps. Two pieces of sheet copper $\frac{1}{16}$ inch thick, the width of the vise jaws one way and about six inches the other, will answer to make the copper clamps. The first operation is to soften them by heating them to a low red heat and dipping them in cold water or brine. We then take one of the pieces of copper and grip it in the vise, so that it just covers but does not project below the gripping faces of the vise jaws, and bend the upper part over to fit the outer face of one vise jaw, making it fit the latter neatly all over and hammering it so as to bring the edge nearly sharp. We then take it from the vise and soften it again, and replacing it in the vise, hammer it on the upper surface till the edge is sharply defined. We then operate upon the other piece of copper in like manner, thus producing a pair of clamps that, having sharply defined and not rounded, gripping edges, will hold a thin piece of work above the upper face of the clamps without springing the work out of the vice. The object of having the upper surface of the copper so long is so that it will bend far over the upper surface of the vise jaw, and will not be liable to fall off when the vise jaws are opened.

Since copper clamps become hardened through use, it is necessary, when they become glazed, to soften them.

Lead clamps are usually made as wide as the vise jaws, the part covering the gripping faces of the vise jaws being of the same depth as those faces and about ¾ inch thick, the upper face of each clamp which laps over the upper jaws of the vise well covering the latter and being about ⅛ inch thick. Both lead and copper clamps are apt to get filings and grit embedded in their surfaces; hence on delicate work it is well to either place a piece of rag or cloth over them, or else to use a pair of leather clamps, say ⅛ inch thick, the leather being cut through about half way to cause it to bend over the vise jaws and not fall off when the vise jaws are opened.

Work which presents a small gripping surface to the vise jaws, such for instance as round bolts or spindles, can be held more firmly by lead than by copper clamps, because the work will sink into the lead when the vise is tightened.

VISE WORK—PENING.

The operation termed pening is stretching the skin on one side of work to alter its shape, the principle of which is that, by striking metal with a hammer, the face of the metal struck stretches, and tends to force the work into a circular form, of which the part receiving the effect of the hammer is the outside circle or diameter.

Fig. 250 represents a piece of flat iron, which would, if it

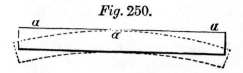

Fig. 250.

were well hammered on the face, *a, a, a*, with the pene of a hammer, alter its form to that denoted by the dotted lines.

Fig. 251 represents a brass which, if struck with a hammer (along its bore at *a*) or other piece of metal for driving it in while fitting, would gradually assume the form denoted by the dotted lines. Fig. 252 represents a rod connected at the end *a* with a double eye

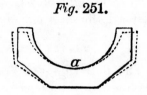

Fig. 251.

and pin, and requiring to descend true so as to fit into the double eye *b*, at the other end; if, therefore, it is pened perpendicularly on the face *c* of the rod, the stretched

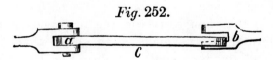

Fig. 252.

skin will throw the end around so that it will come fair with the eye *b*. Connecting rod straps which are a little too wide for the rod ends may be in like manner closed so as to fit by pening the outside of the crown end, or, if too narrow, may be opened by pening the inside of the crown end; but in either case the ends of the strap alter most in consequence of their lengths, and the strap will require refitting between its jaws.

Piston rings may be made of a larger diameter by pening the ring all round on the inside, and there are many other uses to which pening may be used to advantage, such as setting frames, refitting old work, taking the twist out of work, etc.; but it must be borne in mind that if, after a piece of metal has been pened, a cut is taken off it, it will return to its original shape, as the effects of the pening do not extend more than $\frac{1}{64}$ of an inch in depth. When, therefore, a brass or other work requiring to be bored is driven in and out by a piece of metal or a hammer, it stretches the skin; and when the brass is bored, the stretched skin being cut away, it assumes its original shape and hence becomes slack or loose in the strap or block. A light hammer having a round pene should be used; and light blows should be employed for pening, as they are the most effective.

FITTING BRASSES TO THEIR BOXES.

The pattern for a brass which is hexagonal upon the bottom or bedding part should not be made of exactly the same shape as the hexagonal part of the box upon which it beds, because the brass, in casting, shrinks in the direction of the diameter of the bore to such an extent as seriously to alter the angles of the bottom of the brass as compared with the angles on the bottom of the pattern. To compensate for this change of form, the angles on the sides of the pattern should be made more obtuse than those on the sides of the box, as described in Fig. 253, the dotted lines being the angle of the box. The shrinkage referred to is not merely that due to the contraction of the metal in cooling, but is an alteration of form which takes place in all castings of more or less segmental circular form, especially in the case of light castings. In castings of 4 inches or less diameter, the rapping (given by the moulder to the pattern to loosen it in the sand, so as to be able to extract the pattern without damaging the mould) is about equal to this alteration of form; but in larger castings an allowance must be made for it.

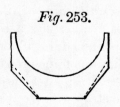

Fig. 253.

In fitting a brass to its box, first fit the sides of the brass to the box, keeping them at an equal angle to the joint or top face of the brass, so as to let the brass down evenly and not with one side or one bevel lower than the other. To find if the brass is level, use inside calipers as a gauge, applied from the top face of the brass to the top face of the box. When the brass is let down so that it approaches the bottom of the box, rub upon the bed of the box a coating of marking; and then upon the end of each bevel, and upon the bottom and near each corner (of the bevels and bottom), place some small pellets of red lead, mixed stiffly; then when the brass is driven home upon its bed and again taken out, the pellets of red lead will adhere to the box because of the marking, and (by their respective thicknesses) denote how nearly the angles or bevels of the brass fit to the box; because where the brass touches the bed of the box, the pellets will be mashed; but if the pellets are intact, it demonstrates that there is space between the box and the brass. It is obvious that the brass requires chipping in those places where the pellets are crushed, and in proportion to the thickness of the pellets that are the least crushed. The pellets should be removed and replaced each time before driving the brass home, and removed when they appear of even thicknesses, the fitting being completed with marking only. All brasses must be fitted to their boxes more tightly than they are intended to be when finished, because they go in from the process of boring and are consequently an easier fit after than before being bored.

FITTING LINK MOTIONS.

The planing and boring of the link of the die, and of the eccentric rod double eyes being completed, the faces of the links may be filed up to a surface plate. The slot of the link should then be filed out to a gauge of sheet-iron of the proper sweep, the sides of the slot being kept square at all parts with the face of the link: each end of the slot at the termination of the stroke of the die should be eased off a little, so that, when the link and the die are hardened, the latter will not bind hard in the ends of the former, as would otherwise inevitably be the case. The die may then be fitted, to a rather tight fit, to the slot of the link, putting a very light coat of marking upon either or both of them, which will serve as a lubricant to prevent them from cutting, and will show the high spots upon both the link and the die, which spots must be eased off until the die fits to a working fit, providing the link and die are not intended to be hardened. If, however, they are to be hardened, the die must be made of a somewhat easier fit to allow for the expansion of the metal which takes place in hardening. To fit the double eyes (that is, the eccentric rod ends) upon the link (or quadrant), a bolt and washer should be provided, the pin being a fit to the hole in the eye and to the hole in the washer, the head of the pin and the washer being the finished diameter of the outside of the eye. The end of the pin is passed through one side of the eye, then through the washer, and then through the other side of the eye.

The underneath faces of the pin and washer will, if revolved by hand, mark the two faces (against which they bear) true with the hole of the double eye; and when those faces are finished, the pin may be turned end for end, and the other two faces trued in the same manner. The object of making the head of the pin and the washer of the same diameter as the double eye is that they may be used as gauges to which to file up the outside of the double eye, for which purpose they should be hardened so that the file will not cut them. The double eyes being filed to fit the link, the washer (having been used, as above described, as a gauge to keep the faces true to the hole) must then be clamped to the link, care being taken to make the hole of the link as true as possible with the hole of the double eye, and to slacken the bolt of the clamp if the double eye requires moving to come fair with the hole in the link. If the clamp were not slacked, striking the double eye to move it would probably spring one jaw out of true with the other. A hand reamer may be passed through the double eye, taking out a light cut, and thus making the holes through the link and double eye parallel and quite true with each other.

If, after the link has been hardened, the die is too tight a fit, place oil and fine emery in the slot, put the die in its place and (with a piece of wood through the hole of the die) force it back and forth from end to end of the slot, or in such parts only as it may be too tight; this will grind out the tight places. If there is no fine emery at hand, crush some coarse emery, using a hammer and a block of iron. If the link is tight at the extreme ends, as is sometimes the case, a piece of flat copper shaped and used as a file may be used with grain emery and oil to grind out such ends. If, however, the link has altered so much as to make the grinding a long and tedious operation, it may be opened by placing a bolt and nut in such a position in the slot that the head of the bolt will rest against one side and the end face of the nut against the other side of the slot; the head of the bolt should then be held stationary with a wrench or spanner, and the nut, being unscrewed, will force open the link. Another method is to take two keys, such as connecting-rod keys, both having an equal amount of taper on them, and place them in the slot of the link with their edges bearing against each, and with the heads of the keys on opposite sides of the link. The operation is to place a hammer against the head of one key, to prevent it from driving out of the link, and to drive in the other key. The advantage of this method over the screw and nut is this: The link will spring considerably before it will alter its form, so that, when applying the bolt and nut, it is difficult, in the second operation (providing the first has not effected the desired opening of the link), to find exactly how much to unscrew the nut. In using the keys, however, lines may be drawn across the keys to denote exactly how far they were driven in during the first operation, which lines will guide the judgment as to how far to drive them in the second operation. If a link opens, that is, if the slot becomes wider during the process of hardening, it may be closed by clamping, or even by a strong vise.

FITTING CYLINDERS.

A casual cylinder or pair of cylinders (there being no templates for marking the holes, etc.) should be fitted up as follows : If that part of the cylinder cover which fits into the cylinder has a portion cut away to give room for the steam to enter (as is usually the case), mark a line across the inside flange of the cover, parallel to the part cut away, and then scribe each end of the line across the edge of the flange. Then mark a similar line across the cylinder end, parallel to the steam port where it enters the cylinder, and scribe each end of this line across the cylinder flange, so that, when the cylinder cover is placed into the cylinder and the lines on the flanges of the cylinder and the cover are placed parallel to each other, the piece cut away on the cover will stand exactly opposite to the steam port, as it is intended to do. The cover may then be clamped to the cylinder, and holes of the requisite size for the tap (the tapping holes, as they are commonly called) may be drilled through the cover and the requisite depth into the cylinder at the same time. Concerning the correct size of a tapping hole in cast-iron, as compared to the tap, there is much difference of opinion and practice. On the one hand, it is claimed that the size of the tapping hole should be such as to permit of a full thread when it is tapped ; on the other hand, it is claimed that two-thirds or even one-half of a full thread is all that is necessary in holes in cast-iron, because such a thread is, it is claimed,

VISE-WORK—FITTING CYLINDERS.

equally as strong as a full one, and much easier to tap. In cases where it is not necessary for the thread to be steam-tight and where the length of the thread is greater by at least ⅛ inch than the diameter of the bolt or stud, three-quarters of a full thread is all that is necessary, and can be tapped with much less labor than would be the case if the hole was small enough to admit of a full thread, partly because of the diminished duty performed by the tap, and partly because the oil (which should always be freely supplied to a tap) obtains so much more free access to the cutting edges of the tap. If a long tap is employed to cut a three-quarter full thread, it may be wound continuously down the hole, without requiring to be turned backwards at every revolution or so of the tap, to free it from the tap cuttings or shavings, as would be necessary in case a full thread was being cut. The saving of time in consequence of this advantage is equal to at least 50 per cent. in favor of the three quarter full thread.

The cylinder covers must, after being drilled, as above, be taken from the cylinder, and the clearing drill put through the holes already drilled so that they will admit the bolts or studs, the clearing holes being made $\frac{1}{16}$ inch larger than the diameter of the bolts or studs. The steam-chests may be either clamped to the cylinder, and tapping holes drilled through it and the cylinder (the same as done in the case of the covers) or it may have its clearing holes drilled in it while so clamped, care being taken to let the point of the drill enter deep enough to pass completely through the steam-chest, and into the cylinder deep enough to cut or drill a countersink nearly or quite equal to the diameter of the drill. If, however, the steam-chest is already drilled, it may be set upon the cylinder, and the holes marked on the cylinder face by a scriber, or by the end of a piece of wood or of a bolt, which end may be made either conical or flat for the purpose, marking being placed upon it; so that, by putting it through the hole of the chest, permitting it to rest upon the cylinder face (which may be chalked so as to show the marks plainly), and then revolving it with the hand, it will mark the cylinder face. This plan is generally resorted to when the holes in the chest are too deep to permit of being scribed. To true the back face, round a hole against which face the bolt head or the face of the nut may bed (in cases where such facing cannot be done by a pin countersink or a cutter used in a machine), the appliance shown in Fig. 254 may be employed, a being a pin provided with a slot at one end to

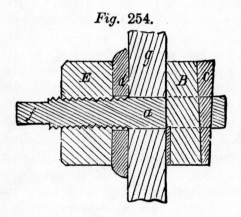

Fig. 254.

admit the cutter B, which is held fast by the key C, and is also provided with a square end f, by which it may be turned or revolved by means of a wrench, and with a thread to receive the nut E, d being a washer; so that, by screwing up the nut E, the cutting edges of the cutter are forced against the cylinder g, and will, when revolved, cut the face against which they are forced, true with the hole in the cylinder through which the pin a is passed.

To fit the cylinder cover joint, put marking on the joint face of the cover; put the cover into its place on the cylinder face; then, in order to discover how much the faces are out of true, strike the outside of the cover on one side of its diameter, and then the other, alternately, with the fist; and if the faces at any point are open, they will strike each other with a blow the sound of which will clearly indicate to what extent they are out of true ; if much, the cover may be removed and the high parts rough-filed to any extent the judgment may indicate; if, however, when the cover is struck, the faces give no sound of striking, smooth filing will answer. When the faces mark nearly all over, the high spots may be eased with the scraper until the surfaces are sufficiently close that a light coating of marking will mark them all over, when they may be ground together as follows: Place on the cylinder face grain emery and oil, and then put the cover on. Fasten to the cover a lever, and then place sufficient weight upon the cover to leave it capable of being conveniently moved by means of the lever (which should project on both sides of the cover). The cover must not be revolved all in one direction, or the emery will cut grooves in the face, but must be moved back and forth while it is being revolved. When the grinding has proceeded until the cover moves smoothly upon the cylinder face, indicating that the emery has worn down and worked out (as it will do) from between the faces, the cover may be removed ; and if the grinding appears equal and of one shade of color all over the faces, the emery may be wiped off them, and the cover replaced and revolved back and forth as before, which will cause the faces to polish each other, removing all traces of the emery, and showing plainly the slightest defect in the surfaces. If, however, the first grinding is not sufficient (as is generally the case), oil and emery must be again supplied, and the grinding continued as in the first instance. The cover of an 18 inch cylinder, even if it is much out, may be made of a steam-tight fit by this process in about half an hour.

It is obvious that, in the case of a large cylinder cover, such as are used for marine purposes, the hand will not strike a sufficient blow to indicate how much the faces, before fitting, are out of true, and a block of wood and a hammer must be employed instead.

The next operation is to cut out the cylinder ports to their requisite dimensions.

In facing up the valve faces, the surface plate may, in like manner, be struck on its opposite corners, or a pressure may be placed on them by the hands to ascertain if the surface plate will rock, and to what extent. If it rocks at all, a rough file should be employed to file away the high parts of the face; if it does not rock, a smooth file should be employed to take out the tool marks, the filing being continued until a light coat of marking on the surface plate will mark the cylinder face all over, when the scraper may be applied to finish it. The slide valve itself may then be surfaced and scraped to the surface plate, and then placed upon the cylinder face, and the valve and cylinder face scraped together.

The joint of the steam-chest may be made by filing the planed surfaces to a straight edge, and placing between the chest and its seat on the one hand, and the cover and the chest on the other hand, a lining of very thin softened sheet copper, which plan is generally adopted on cylinders for locomotives.

In cases where a number of cylinders of similar sizes are made, the whole of the marking off, and much other work, may be saved by the employment of gauges, etc.

For drilling the cylinder covers and the tapping holes in the cylinder, the following system is probably the most advantageous: The flanges of the cylinder covers are turned all of one diameter, and a ring is made, the inside diameter of which is, say, an inch smaller than the bore of the cylinder; and its outside diameter is, say, an inch larger than the diameter of the cover. On the outside of the ring is a projecting flange which fits on the cover, and which is provided with holes, the positions of which correspond with the required positions of the holes in the cover and cylinder; the diameter of these holes (in the ring, or template, as it is termed) is at least one quarter inch larger than the clearing holes in the cylinder are required to be. Into the holes of the template are fitted two bushes, one having in its centre a hole of the size necessary for the tapping drill, the other a hole the size of the clearing drill; both these bushes are provided with a handle by which to lift them in and out of the template, and both are hardened to prevent the drill from cutting them, or the borings of the drill from gradually wearing their holes larger. The operation is to place the cover on the cylinder and the template upon the cover, and to clamp them together, taking care that both cover and template are in their proper positions, the latter having a flat place or deep line across a segment of its circumference, which is placed in line with the part cut away on the inside of the cover to give free ingress to the steam, and the cover being placed in the cylinder, so that the part so cut away will be opposite to the port in the cylinder, by which means the holes in the covers will all stand in the same relative position to any definite part of the cylinder, as, say, to the top or bottom, or to the steam-port, which is sometimes of great importance (so as to enable the wrench to be applied to some particular nut, and prevent the latter from coming into contact with a projecting part of the frame or other obstacle). The bush, having a hole in it of the size of the clearance hole, is the one first used, the drill (the clearance size) is passed through the bush, which guides it while it drills through the cover, and the point cuts a countersink in the cylinder face. The clearing holes are drilled all round the cover, and the bush, having the tapping size hole in it, is then brought into requisition, the tapping drill being placed in the drilling machine, and the tapping holes drilled in the cylinder flange, the bush serving as a guide to the drill, thus causing the holes in the cover and those in the cylinder to be quite true with each other. A similar template and bush is provided for drilling the holes in the steam-chest face on the cylinder, and in the steam-chest itself. While, however, the cylinder is in position to have the holes for the steam-chest studs drilled, the cylinder ports may be cut as follows, which method was introduced in 1867, with marked success, by Mr. John Nichols, who was then manager of the Grant Locomotive Works at Paterson, N. J.: The holes in the steam-chest face of the cylinder being drilled and tapped, a false face or plate is bolted thereon, which plate is provided with false ports or slots, about three-eighths of an inch wider and three-fourths of an inch longer than the finished width and length of the steam-ports in the cylinder (which excess in width and length is to allow for the thickness of the die). Into these false ports or slots is fitted a die, to slide (a good fit) from end to end of the slots. Through this die is a hole the diameter of which is that of the required finished width of the steam-ports of the cylinder. Into the hole of the die is fitted a reamer, with cutting edges on its end face and running about an inch up its sides, terminating in the plain round parallel body of the reamer, whose length is rather more than the depth of the die. The operation is to place the reamer in the drilling machine, taking care that it runs true, place the die in one end of the port, and then wind the reamer down through the die so that it will cut its way through the port of the cylinder at one end; the spindle driving the drill is then wound along. The reamer thus carries the die with it, the slot in the false face acting as a guide to the die.

In the case of the exhaust port, only one side is cut out at a time. It is obvious that, in order to perform the above operation, the drilling machine must either have a sliding head or a sliding table, the sliding head being preferable.

The end of the slot at which the die must be placed when the reamer is wound down through the die and cylinder port, that is to say, the end of the port at which the operation of cutting it must be commenced, depends solely on which side of the port in the cylinder requires most metal to be cut off, since the reamer or cutter, as it may be more properly termed, must cut underneath the heaviest cut, so that the heaviest cut will be forcing the reamer back. The reason for the necessity of observing these conditions, as to the depth of cut and direction of cutter travel, is that the pressure of the cut upon the reamer is in a direction to force the reamer forward and into its cut on one side, and backward and away from its cut on the other side, the side having the most cut exerting the most pressure. If, therefore, the cutter is fed in such a direction that this pressure is the one tending to force the cutter forward, the cutter will spring forward a trifle, the teeth of the cutter taking in consequence a deep cut, and, springing more as the cut deepens, terminate in a pressure which breaks the teeth out of the cutter. If, however, the side exerting the most pressure upon the reamer is always made the one forcing the cutter back, by reason of the direction in which the cutter is travelled to its cut, the reamer in springing away from the undue pressure will also spring away from its cut, and will not, therefore, rip in or break, as in the former case.

In cutting out the exhaust port, only one side, in consequence of its extreme width, may be cut at one operation; hence there are two of the slots provided in the false plate or template for the exhaust port. The cutter must, in this case, perform its cut so that the pressure of the cut is in a direction to force the cutter backwards from its cut. The time required to cut out the ports of an ordinary locomotive cylinder, by the above appliance, is thirty minutes, the operation making them as true, parallel, and square as can possibly be desired.

In order to tap the holes in the cylinder heads and steam-chest seat on the cylinder true, without requiring the workman to apply the square, a long tap and a guide is employed for holding the guide to the cylinder face. If the end cylinder faces have a projecting ring on them (so as to leave a small surface to make the joint), the guide may be cut away on its bottom face to fit the projection, so that if the guide is held against the projection, while the guide is bolted fast, the hole in the guide through which the tap passes will stand true (both ways) of itself, to the hole to be tapped in the cylinder. In the case, however, of there being no projection of the kind mentioned, as, for instance, when tapping the holes in the seat for the steam-chest, the guide will require adjusting sideways, by the eye. The distance, however, of the holes in the guide being the same from centre to centre as the distance from centre to centre of the holes to be tapped, insures, without any setting, that the holes tapped are true with each other one way.

The saving of time and labor effected by means of the employment of this system and its appliances is much greater than might be supposed at first sight; it may, however, be appreciated when it is stated that, under it, three pairs of locomotive cylinders have been fitted up by one skilled workman and one assistant in seven and a half days, the work done to each pair being the holes, amounting to 200, drilled, and those for the cylinder covers, cylinder cocks, steam-chests, steam-pipes, and exhaust-pipes, tapped; the steam and exhaust ports cut out, and the faces and those of the slide valves scraped up, the cylinder end and cover faces filed, scraped, and ground up steamtight, the steam-chest seat faces filed up true to a straight edge, the seat for the steam and exhaust ports faced out with the cutter, all necessary bolts and studs put in, the cylinders bolted together, their bores being set true with each other, and the whole turned out so that the cylinders were complete and ready to bolt to the engine frames.

For screwing the studs into a cylinder the following tool should be used: a piece of iron, say four inches long, should have one end cut square for a wrench to fit, and in the other end a hole, tapped clear down to its bottom and fitting the thread of the stud loosely; the stud will then bottom in it when being screwed in, while, when the stud is home in the cylinder, the tool will leave it easily without unscrewing it again.

SCRAPED SURFACES.

There are many who believe that surfaces properly planed are sufficiently true for all ordinary practical purposes; but if those persons were to apply a surface plate to a well-planed surface a foot square, or to a common connecting rod, key and gib, as they usually leave the planer, they would be effectually cured of the fallacy of their opinion upon this matter.

It is impossible to hold a piece of work sufficiently firm that it can be cut by a machine tool without springing it; and though in stout work having a fair bedding beneath the clamping bolts, and in work in which the pressure referred to is sustained in a direction to directly compress the metal of the work, the amount of this spring may be almost imperceptible, still, in light and in a great deal of other work, the amount of spring due to the pressure of holding the work is sufficiently great to throw it out of true. Another of the reasons for the necessity of using surface plates is, that all work alters its form from having its surface skin removed, as will be hereafter explained. All working flat surfaces should be surfaced to a surface plate, whether they are intended to be finished with a file or a scraper. Scrapers are not intended for use as tools to take off a quantity of metal, but for the purpose of making the work very true, being used by itself or in conjunction with a file to ease away the high spots; not because it is impossible with a file of even sweep and flat cross surface to file true, but because it is a quicker and easier method of obtaining a flat surface, and one that is absolutely indispensable to fine work if the file has warped to a sensible degree in the hardening. If, after a piece of work has been planed and the surface plate has been applied, it is found that the surface is somewhat out of true, as is generally the case, it is better to file the work until the surface is true, which process will be quicker than scraping from the commencement.

It is useless to apply a scraper promiscuously over a surface for the purpose of making it appear smooth; for a surface can be got up, so far as smoothness is concerned, far better with a file and emery paper than it can be with a scraper. Fig. A represents a surface finished by a file and emery paper, the surface being so fine that even common paper will scratch it.

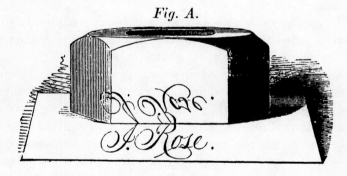

Fig. A.

The proper method of procedure in scraping a flat surface is to first go all over it, leaving the scraper marks as shown in Fig. B.

The second time of going over the surface should leave the marks as shown in Fig. C; while the surface will appear after the third scraping as in Fig. D.

After each scraping we apply the surface plate and rub it well over the work to mark it, giving the surface of the plate a barely perceptible coat of marking, and distributing the same evenly all over with the palm of the hand, so as to detect any grit that may chance to have got into the marking. A piece of *old* rag should be used for wiping the surfaces clean (which is better than either new rag or

Fig. B.

Fig. C.

waste) and great care should be taken that the surfaces have no dust or grit upon them, or it will scratch the surfaces. The surface plate is made to mark the work by being rubbed back and forth upon it, or, if the work is small, it may be taken from the vise and rubbed upon the face of the surface plate. In either case, the high spots upon the face of the work will become very dark, or, if the amount of marking applied was barely sufficient to dull the surface of the plate, the marks will be almost black and will, by continuous rubbing on the plate, become bright.

Here we may observe that small work applied to the

Fig. D.

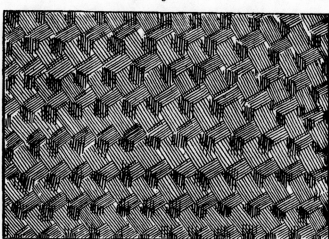

plate should be rubbed at the corners and toward the outer edges of the plate so as to keep the wear of the latter as even as possible, since the middle of the surface of the plate generally suffers the most from use, becoming in time hollow.

The harder the plate bears upon the work the darker the spots, where it touches, will appear; so that the darker the spots the heavier the scraping should be performed. It will be noted that the scraper marks are much smaller and finer at and during the last scrapings, and it may be

VISE-WORK—MAKING SURFACE PLATES.

here remarked that the scrapings are taken very light and in a direction lengthwise of the surface plate marks during the finishing process.

The best form of scraper is that shown in Fig. E, which should be ground down so that the cutting edge does not

Fig. E.

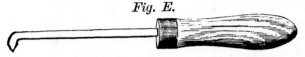

stand more than a quarter of an inch from the body of the tool during the finishing process; for the edge will not cut so smoothly if the cutting end and edge is bent far out. After the scraper is ground, it should be carefully oil-stoned to a smooth edge. For use upon brass, wrought-iron and steel, the cutting edge should be kept moistened with water.

TO MAKE A SURFACE PLATE.

To obtain a true surface plate we must *first* get up three plates, which we will term numbers 1, 2 and 3.

We take No. 1, and placing a true straight edge across it we take one end of the straight edge, between the finger and thumb, and taking care not to place any vertical pressure upon it, we move it sideways back and forth about an inch, to see where on the plate the fulcrum of its movement takes place. If the centre of its movement is at the centre of the plate, then the surface of the plate is rounding—that is, highest in the middle. If the straight edge moves on the plate, first, the most at one end, and then the most at the other end, it demonstrates that the straight edge takes its fulcrum of movement towards the edges of the plate, and hence that the latter is hollow. If, however, the straight edge moves with a shuffling movement, it denotes that the plate is neither rounding nor hollow. The surface, however, may nevertheless be atwist and the straight edge will not detect it, unless two are used—being placed parallel to each other, one at the opposite side of the plate to the other—when, if the straight edges are sufficiently long, the eye, directed across the upper edges of the straight edges, looking at them sideways, will readily detect any twist. Having levelled Nos. 1 and 2 as near as possible by the straight edge, we place marking on one of them and then rub their surfaces well together, to mark them, and then proceed to scrape them until they fit together. We should, however, when putting their faces together for the first time, strike the back of the top plate a sharp blow at each corner with the closed hand, and if the surfaces are atwist at all, a blow distinctly heard will be given by the top plate to the bottom one; the degree of loudness of the blow will also denote how much the surfaces are out of true. It may be thought that the surfaces, being planed, cannot well be atwist. Such, however, is not the case, nor is the twist due to the planing but to the alteration of form which takes place in the work, by reason of the tension of the skin upon the face having been removed. This alteration of form may be reduced to a minimum by slacking back the bolts, plates or jaws, holding the work in the planer, previous to taking the last cut in the planer, since the last cut being a light one, not much pressure is required to hold the work.

Fig. F.

Having scraped Nos. 1 and 2 together, we introduce No. 3 and scrape it to fit No. 2, not scraping the latter at all. We next try Nos. 1 and 3 together, and if they show each other to be rounding, it is proof that No. 1 is rounding to one-half the amount of difference between it and No. 3 as shown in Fig. F.; from which it will be perceived that the two nearest together faces of 1

VISE-WORK—MAKING SURFACE PLATES. 303

and 2 may fit together, one being rounded and the other hollow. No. 2 may then be taken as a gauge whereby to fit No. 3. But if we then take Nos. 1 and 3, and try them together, they will disagree to twice the amount that each of them separately is out—that is to say, twice the amount that No. 1 varied from being a true flat surface.

*We next scrape Nos. 1 and 3 together, taking off, as nearly as our judgment dictates, an equal amount off each with the scraper, and this being complete, we scrape No. 2 to No. 1, and then No. 3 to No. 2, finally bringing back No. 3 to No. 1, as a test, scraping them together, taking an equal amount off each, and when they fit perfectly we fit 2 again to 1, and 3 again to 2, continuing the whole operation until all three plates fit, each the others, perfectly.

Here, however, we may note as follows : the three plates being of the same size, as they should be, and it being necessary to rub their faces together in order to mark where they touch together, the abrasion of one face against the other, and therefore the marks, will not be equal Thus, in Fig. G, plate 1 being moved forward, that part

Fig. G.

of its surface overlapping and denoted by C, and that part of plate 2 denoted by B, are not in contact with a surface, and are not, therefore, being marked by the movement, whereas the whole of the rest of the surfaces of the two plates *are* in contact, and are therefore marking each other during the whole of the movement, the consequence being that those parts of the surfaces which overlap mark more lightly in proportion to the amount of their bearing than the rest of the surfaces, and since it is the lightness or heaviness of the marks which determine how much the parts of contact must be eased by the scraper, it becomes evident that a surface plate cannot be made true to the highest practical attainable degree, unless, having finished the three plates, we introduce a *fourth*, whose size shall be sufficiently smaller that it can be rubbed back and forth upon the plate or plates by which it is surfaced without overlapping at all. The feet upon which a surface plate rests should be planed true, and the plate resting upon the bench should rest equally upon all of the feet, otherwise the plate will deflect from its own weight, the amount of the deflection making a practical difference in a large plate.

TO CUT HARD SAW BLADES.

Mark out on the saw blade the article you require to cut from it and centre-punch it lightly all along the lines, going over and over it until the centre-punch marks, which should not be more than $\frac{1}{16}$ inch apart, are nearly through the blade; then take a hard chisel and nick between and along the centre-punch marks. Then reverse the saw blade, and centre-punch from the other side, and the blade will cut to almost any shape without splitting.

Keys should be made a snug fit and parallel on the sides, having taper on the top and bottom only. They should be fitted to bear even all over, or they spring the work out of true. A very light coat of marking should be used in fitting them, and they should be driven in and out lightly.

TO REFIT LEAKY PLUGS TO THEIR COCKS.

When a cock leaks, be it large or small, it should be refitted as follows, which will take less time than it would to ream or bore out the cock or to turn the plug, unless the latter be very much worn indeed, while in either case the plug will last much longer if refitted, as hereinafter directed, because less metal will be taken off it in the refitting.

VISE-WORK—COCKS AND PLUGS.

After removing the plug from the cock, remove the scale or dirt which will sometimes be found on the larger end, and lightly draw-file, with a smooth file, the plug all over from end to end. If there is a shoulder worn by the cock at the large end of the plug, file the shoulder off even and level. Then carefully clean out the inside of the cock, and apply a very light coat of red marking to the plug, and putting it into the cock press it firmly to its seat, moving it back and forth part of a revolution; then, while it is firmly home to its seat, take hold of the handle end of the plug, and pressing it back and forth at a right angle to its length, note if the front or back end moves in the cock; if it moves at the front or large end, it shows that the plug is binding at the small end, while if it moves at the back or small end, it demonstrates that it binds at the front or large end. In either case, the amount of movement is a guide as to the quantity of metal to be taken off the plug at the requisite end to make it fit the cock along the whole length of its taper bore. The red marking referred to is dry Venetian red and lubricating oil, mixed thickly, a barely perceptible coating being sufficient.

If the plug shows a good deal of movement when tested as above, it will be economical to take it to a lathe, and, being careful to set the taper as required, take a light cut over it. Supposing, however, there is no lathe at hand, or that it is required to do the job by hand, which is, in a majority of cases, the best method, the end of the cock bearing against the plug must be smooth filed, first moving the file round the circumference, and then draw-filing; taking care to take most off at the end of the plug, and less and less as the other end of the plug is approached. The plug should then be tried in the cock again, according to the instructions already given, and the filing and testing process continued until the plug fits perfectly in the cock. In trying the plug to the cock, it will not do to revolve the plug continuously in one direction, for that would cut rings in both the cock and the plug, and spoil the job; the proper plan is to move the plug back and forth at the same time that it is being slowly revolved. As soon as the plug fits the cock from end to end, we may test the cock to see if it is oval or out of round, as follows: First give it a very light coat of red marking, just sufficient, in fact, to well dull the surface, and then insert the plug, press it firmly home, and revolve it as above directed; then remove the plug, and where the plug has been bearing against the surface of the cock, the latter will appear bright. If, then, the bore of the cock appears to be much oval, which will be the case if the amount of surface appearing bright is small, and on opposite sides of the diameter of the bore, these bright spots may be removed with the half-round scraper. Having eased off the high spots as much as deemed sufficient, the cock should be carefully cleaned out (for if any metal scrapings remain they will cut grooves in the plug), and the red marking reapplied, after which the plug may be again applied. If the plug has required much scraping, it will pay to take a half-round smooth file that is well rounding lengthwise of its half round side, so that it will only bear upon the particular teeth required to cut, and selecting the highest spot on the file, by looking down its length, apply that spot to the part of the bore of the cock that has been scraped, draw-filing it sufficient to nearly efface the scraper-marks. The process of scraping and draw-filing should be continued until the cock shows that it bears about evenly all over its bore, when both the plug and the cock will be ready for grinding.

Here, however, it may be as well to remark that in the case of large cocks we may save a little time and insure a good fit by pursuing the following course, and for the given reasons. If a barrel bears all around its water-way only for a distance equal to about $\frac{1}{16}$ of the circumference of the bore, and the plug is true, the cock will be tight, the objection being that it has an insufficiency of wearing surface. It will, however, in such case wear better as the wearing proceeds. There is perhaps the further objection that so small an amount of wearing surface may cause it to abrade. This, however, has nothing to do with our present purpose, which is to save time in the grinding, insure a good fit, and, at the same time, ample wearing surface. Our plug and barrel being fitted as directed, we may take a smooth file and ease very lightly away all parts of the barrel, save and except to within say ⅜ inch around the water or steam way. The amount taken off must be very small—indeed, just sufficient, in fact, to ease it from bearing hard against the plug, and the result will be that the grinding will bed the barrel all over to the plug, and insure that the metal around the water or steam-way on the barrel shall be a good fit, and hence that the cock be tight. In the case of large cocks, the barrel and the plug should stand vertical during both the final trials and the grinding process.

The best material to use for the grinding apparatus is the red burnt sand from the core of a brass casting, which should be sifted through fine gauze, and riddled on the work from a box made of say a piece of 1½ pipe 4 inches long, closed at one end, and having fine gauze instead of a lid. Both the barrel and the plug should be wiped clean and free from filings, etc., before the sand is applied; the inside of the barrel should be wetted, and the plug dipped in water, the sand being sifted, a light coat, evenly over the barrel and the plug. The plug must then be inserted in the barrel without being revolved at all till it is home to its seat, when it should be pressed firmly home, and operated back and forth while being slowly revolved. It should also be occasionally taken a little way out from the barrel and immediately pressed back to its seat and revolved as before, which will spread the sand evenly over the surfaces and prevent it from cutting rings in either the barrel or the plug. This process of grinding may be repeated, with fresh applications of sand, several times, when the sand may be washed clean from the barrel and the plug, both of them wiped comparatively dry and clean, and the plug be reinserted in the barrel, and revolved, as before, a few revolutions; then take it out, wipe it dry, reinsert and revolve it again, after which an examination of the barrel and plug will disclose how closely they fit together, the parts that bind the hardest being of the deepest color. If, after the test made subsequent to the first grinding operation, the plug does not show to be a good, even fit, it will pay to ease away the high parts with a smooth file, and repeat afterwards the grinding and testing operations.

To finish the grinding, we proceed as follows: give the plug a light coat of sand and water, press it firmly to its seat, and move it back and forth while revolving it, lift it out a little from its seat at about every fourth movement, and when the sand has ground down and worked out, remove the plug, and smear over it evenly with the fingers the ground sand that has accumulated on the ends of the plug and barrel; then replace it in the barrel, and revolve as before until the plug moves smoothly in the barrel, bearing in mind that if at any time the plug, while being revolved in the barrel, makes a jarring or grating sound, it is cutting or abrading from being too dry. Finally, wipe both the barrel and the plug clean and dry, and revolve as before until the surfaces assume a rich brown, smooth and glossy appearance, showing very plainly the exact nature of the fit. Then apply a little tallow, and the job is complete and perfect.

REFITTING WORK BY SHRINKING IT.

For closing long holes, boxes, etc., the water process may be employed, as represented in Fig. H. *a a* is the section of a wrought-iron square box or tube, which is supposed to be made red hot and placed suddenly in the water, B, from its end C to the point D, and held there until the end submerged is cold; the result is that the metal in the water, from C to D, contracts or shrinks in diameter, and compresses the hot metal immediately above the water-line,

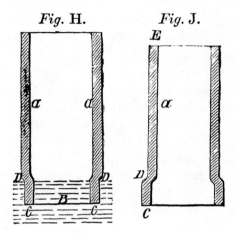

Fig. H.   *Fig.* J.

as the small cone at D denotes. If, then, the box or tube is slowly immersed in the water, this action of compression is carried all the way up from that point, and its form, when cold, will be as described in Fig. J, that part from C to D maintaining its original size, and the remainder being smaller.

It must then be reheated and suddenly immersed from the end E nearly to D, held there until it is cold, and then slowly lowered in the water, as before, which will contract the part from D to C, making the entire length parallel but smaller, both in diameter and bore, than before it was thus operated upon.

Small holes to be reduced in bore by this process should be filled with fire clay, and the faces nearly or wholly covered with the same substance, so that the water will first cool the circumference, and the circumference will, in contracting, force inwards the metal round the hole, which is prevented from cooling so quickly by the clay, and therefore gives way to the compressing force of the outside and cooler metal.

This principle may be made use of for numerous purposes, as for reducing diameters of the tires of wheels, reducing the size of wrought-iron bands, or for closing in connecting rod straps to refit them to the block end, the mode of operation for which is, in the case of a rod whose strap is held by bolts running through the block and strap, to bolt the strap on the rod to prevent it from warping, to then heat the back of the strap, and (holding the rod in a vertical position) submerge the back of the strap in water to nearly one-half its thickness.

If the bolts are not worn in the holes, or if the strap is one having a gib and key, they may be merely put into their places without placing the strap on the rod. Even a plain piece of iron shrinks by being heated and plunged into water, but only to a slight degree, and the operation cannot be successfully repeated. Eccentric rods, which require to be shortened, say $\frac{1}{64}$ of an inch, may be operated on in this manner, in which case care must be taken to immerse them evenly so as not to warp them.

Much labor and expense may often be saved by employing the principles of expansion and contraction to refit work. For instance, suppose a bolt has worn loose: the bolt may be hardened by the common prussiate of potash process, which will cause it to increase in size, both in length and diameter. The hole may be also hardened in the same way, which will decrease its diameter; and if the decrease is more than necessary, the hole may be ground or "lapped" out by means of a lap. Only about $\frac{1}{64}$ of an inch of shrinkage can be obtained on a hole and bolt by hardening, which, however, is highly advantageous when it is sufficient, because both the hole and the bolt will wear longer for being hardened.

Set screws should be made of steel, the end being cupped and the thread at the end being chamfered off, and that end being hardened.

To put a fixed feather in a shaft, cut out the seat for the feather and then take a chisel-set, and bulge out the sides of the recess. The feather (which should have the part intended to be set in the shaft slightly dovetailed, the largest part of the dovetail resting on the bottom of the recess) should then be put in its place, and the metal of the shaft should be closed around the feather, a set being used for the purpose; thus the feather will be riveted in its place, while the surface of the shaft will retain its roundness, because of the sides of the recess having been bulged as directed. After the feather is fastened, the surface of the shaft may be filed smooth and even.

Standing bolts or studs which are placed in a position liable to corrode them should have the standing ends $\frac{1}{8}$ inch larger than the nut end, and the plain part should be square. By this means a wrench may be applied to extract them when necessary, and the stud is not so liable to break off in consequence of weakness, at the junction of the thread, and the plain part where the groove to relieve the termination of the thread is cut.

Bolts that have become corroded in their holes so that they are liable to twist or break off in extracting them, should be well warmed by a red hot nut or washer, because the strength of the stud increases by being heated up to about 400° Fahr., and, therefore, studs which readily twist off when cold will unscrew when heated to about that temperature.

Nuts upon standing bolts in the smoke-boxes of locomotives or in similar positions, which have become so corroded as to endanger twisting the bolt off, should be cut through with a cape or cross-cut chisel, thus saving the stud at the expense of the nut. The split must be cut from the outside end face to the bedding face of the nut. To ease a nut that is a little too tight in the thread, screw it upon the bolt, and, resting it upon a block of iron, strike the upper side with a hammer, turning the nut so that not more than two blows will fall upon one face at a time, and striking light blows towards the last part of the operation.

To ease a nut that is too tight to its place to unscrew, so that an ordinary wrench will not move it, strike a few sharp blows on its end face; then, holding a dull-edged chisel firmly across the chamfer of the nut, strike the chisel-head a few sharp blows. Another plan is to put the full force of the wrench upon the nut, and strike at the same time the nut a few sharp blows upon its end face. If the nut is a large one, a piece of tube applied to the end of the wrench, together with blows struck on the end face, will often suffice to start the nut, especially if the nut is warmed.

Brass castings under 12 inches in size shrink about $\frac{1}{8}$ inch to a foot in cooling in the mould. Large castings shrink about $\frac{3}{16}$ inch.

To prevent air-holes in copper castings they should be moulded in green sand moulds, using as a flux $1\frac{1}{2}$ lbs. of zinc to every 100 lbs. of copper. Pure copper will not cast without honey-combing.

In casting iron on wrought-iron or steel spindles, the metal should be poured endwise of the mould, letting the cast metal cover the spindle an inch longer on the top end than is necessary. Thus the air-holes, if any, will form in the extra inch of length and may be cut off with it in the lathe.

Iron castings shrink in the mould about $\frac{1}{10}$ of an inch to the foot. The shrinkage sideways and endways of a casting 4 inches or less in size is compensated for by the shake in the sand given by the moulder to the pattern in order to extract it from the mould.

In very small castings requiring to be of correct size, allowance should be made in the pattern for the shake of the pattern of the sand, thus: A pattern to cast an inch cube will require to be made $\frac{1}{32}$ inch less than an inch endwise and sideways.

STEAM AND WATER JOINTS.

The best joint of any is the ground or scraped joint, but as this is for many purposes too costly, the following, based upon a lengthy practical experience, will be found reliable:

In fitting flanges to boilers or flanges together, be careful that the closest contact is around the hole. From the inside of the bolt-hole to the outside of a flange should be eased away a trifle whenever the bolts are standing bolts or are not liable to leak.

Red lead joints.—Take white lead ground in oil, and mix with it dry red lead sufficient to make it spread with a steel blade without sticking to it. Thorough bray the mixture by well hammering it with a hand hammer.

Gauze joints for high temperatures are easily made and are lasting. Fine iron gauze is cut to entirely cover the joint, and a coating of red lead mixed as above is laid over it. If the surfaces of the joint are very uneven, the gauze may be doubled.

For joints where hot or cold water are concerned canvas or duck may be used, having on it a coating of red lead mixed as above.

For ordinary joints combination rubber is an excellent material. It consists of alternate layers of rubber and canvas. When making such a joint, one of the surfaces of the rubber should be chalked to prevent the rubber from tearing if it becomes necessary to break the joint.

Rust joint.—Iron turnings 100 lbs., sal ammonia 1 lb., sulphur $\frac{1}{2}$ lb. If the joint is required to set very quick add $\frac{1}{4}$ lb. more sal ammonia. The whole should be thoroughly mixed and just covered with water.

For very high temperatures, a dry heat asbestos board, soaked with thin red lead and oil, makes the best joint.

CHAPTER XV.

FITTING CONNECTING RODS.

The planing work on a connecting rod being complete, the first thing for the fitter to do is to mark off the keyways, the bolt holes (if there are any), the holes for the set screws, the oil holes, etc., so as to have the drilling completed before the straps or rod ends are filed up, because drills leave a burr where they come through the metal, and because the clamps, which hold the work while it is being drilled, are apt to leave marks upon it. The holes should then be tapped, when the rod will be ready for the file. The faces of the rod whereon the straps fit should then be surfaced with a surface plate, and made quite square with the broad faces of the rod, parallel crosswise with each other, and a little taper with each other in the length. The strap should be made narrower between its jaws than the width of the rod end, so as to require to spring open when placed upon the rod end if the brasses are not in their places. The inside faces of the jaws of the strap must be made quite square with the side faces, so that, when the strap is placed upon the rod end, the latter faces of the strap will not spring out true with the broad faces of the rod end. The rod end must have a light coating of marking rubbed over it, and the strap moved back and forth on it, so that the rod end serves as a gauge and surfacing-block to the strap.

If, when the strap is on its place, its side faces are uneven with the side faces of the rod end, as shown in Fig. 255 (which is a sectional view of a strap and rod end, *a* being the rod end, and B B the jaws of the strap), either one or

FITTING CONNECTING RODS. 315

both of the inside side faces of the strap require filing in the direction denoted by the dotted lines, because it is only in consequence of the inside faces not being square with the outside faces that this twist occurs. The keyways in the strap and rod end should be filed out together, that is, while the strap is on its place and secured by being clamped or bolted. If the strap is one held to the rod by a gib and key, the width, from the end of the rod to the crown of the strap when it is placed in position to cut or file out the keyway, should be that of the extreme width of the brasses when the joint of the brasses is close, less the amount of taper there is on the key.

Fig. 255.

The strap, after being fitted to the rod, should be clamped to the rod end, the keyway in the strap and in the rod being placed fair with each other, before the clamp is tightened, for moving the strap after it is clamped will spring it out of true, so that, when the clamp is taken off, the keyways will not be true with each other though they were filed true. In driving in the keys and gibs to fit them, be careful to put a light coat of marking on them, not only to show where they bind but to prevent them from seizing in the keyway. The key and gib placed edgewise together should be parallel on the outside edges, and the keyway should be parallel both edgewise and across its width. A thin sheet-iron gauge is better to measure the thickness of the keyway than inside calipers are, and the same gauge will do to plane the key and gib, leaving them a little full in the thickness. The keyways should be surfaced with a surface plate, its breadth being equal to that of the gib and key together when the head of the key is even with the head of the gib; then when the keyway is finished, and the strap is placed in its intended position on the end of the rod, the strap will have moved back from off the rod end for a distance equal to the amount of the taper on the key, so that there will be the requisite amount of draw on the keyway of the strap on the one side and on the keyway of the rod on the other side, while the key will at the same time come through the strap to its required distance. The faces of the rod end, whereon the jaws of the strap fit, having been made (as directed) a little taper, and the strap allowed (as described) a little spring, the rod end will enter the strap somewhat easily, and tighten as it passes up the strap, so that, when quite up, the strap will fit a little tighter than it is intended, when finished, to do. When the strap is fitted and keyed to the rod, a light cut should be taken off the faces of the rod and strap while they are together, the bolts of a bolt rod being sufficient to hold the strap for that purpose; but in the case of a gib and key, a piece of wood should be placed between the rod end and the crown of the strap, that is, in the space intended to be filled by the brasses, and the wood keyed up so as to lock the strap on the rod while the faces of the rod and strap are planed This being complete, the strap is ready to receive the brasses. The bottom or back brass must be made to a tight fit, so as to spring the strap open sufficiently to make it fit the rod end as easily as required; thus both the brass and the strap will be closely fitted. The top brass must be fitted to the strap while the bottom brass is in its place in the strap, and must be made to fit the strap without being so tight as to spring it open. The corners of both brasses where they fit the corners of the strap should be eased away with the edge of a half-round file, so that they will not destroy the corners of the strap (when the brasses are being driven in and out to fit), which would make the strap appear to be a bad fit on the rod.

While fitting the top brass, it is necessary to try the strap on the rod end (the brasses being in their places) at intervals, so as not to take any more off the top brass than is necessary to let the strap fit the rod end. As a guide, when fitting the brasses to the strap, the calipers may be set to

FITTING CONNECTING RODS.

the width of the rod end where the strap fits, and applied to the strap when the brasses are driven in to fit. The gib and key must, when placed together edgeways, be quite parallel in their total breadth, so that they will fit properly against each other and against the keyway in the rod end and the strap. When setting the gauge for the size to which the brasses are to be planed, place the strap on the rod end to get the correct size, for the strap is narrower (between its jaws) when it is off than when it is on the rod, because of the spring. In bedding the back brass to the strap, let

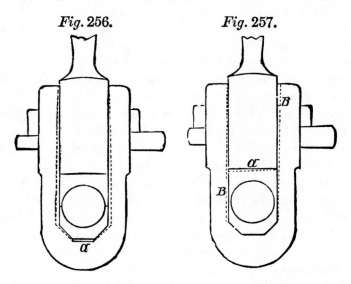

Fig. 256.   *Fig.* 257.

it bear the hardest, if anything, upon the crown, for if the bevels of the brass should keep the crown from bedding, the strap would spring away from the rod end, in spite of the gib (or the bolts, if there are any), when the key is driven home, as illustrated in Fig. 256.

If the back brass does not bed down upon the crown a of the strap, the latter will spring away from the block end of the rod and from the brasses on the sides, and will assume the shape denoted by the dotted lines. Should the top brass not bed properly against the rod end, the strap will spring as described in Fig. 257.

The dotted line *a* is the back of the brass, supposed to bed improperly against the rod end, as shown; the dotted lines B B denote the manner in which the strap would, in consequence, spring away from the rod end when the key was driven home. If the brasses fail to fit properly against the rod end or strap, in the direction of the breadth of the strap, it will spring out of line, as described in Fig. 258. which is a sectional view of a connecting rod end. C is the strap, D is the rod end, and B B are the brasses, the top one of which, if it did not fit square against the rod end (but on one side only), as represented by the line *a*, would spring the strap out of true with the rod end, in the direction of the dotted lines. The strap is, by reason of its shape, very susceptible to spring; and unless the brasses, or even the gib and key, are quite square and fit well, it is certain to spring out of true. The brasses should be a fit on the journal when they are "brass and brass," that is, the joint of the two brasses close together, so as to take the pressure of the key, which thus locks the strap and brasses to the rod end, and prevents them from moving, or working, as it is called, when the rod is in action; especially is this necessary in straps having a gib and key to hold them to their places, because, if the joint of the brasses is not close, the key cannot be driven home tightly, and hence there is nothing to lock the strap firmly to its place. If, however, the strap is held to its place by bolts, it is not so imperative to keep the joint of the brasses close together, although it is far preferable to do so, especially in the case of fast-running engines, not only on account of the assistance lent by the key to hold the strap firmly, but also because it holds the brasses firmly, and the key cannot bind the brasses too tightly to the journal, even though the key be driven tightly

Fig. 258.

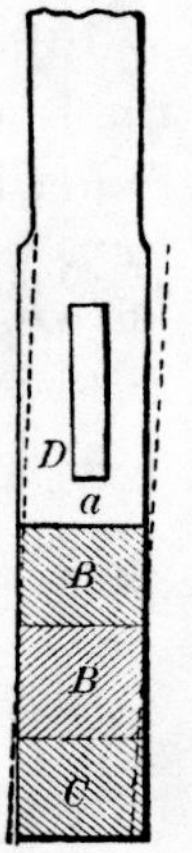

FITTING CONNECTING RODS.   319

home, so as to assist the set screw in preventing it from slacking back.

The brasses should be left a little too tight in the strap before boring, because they invariably shrink or go in a little sideways from being bored, as do all brasses, large or small, even if bored before any other work has been done on them.

For driving the brasses in and out of the strap to fit them, use a piece of hard wood to strike on so as not to stretch the skin of the brass and alter its form, as already explained in the remarks on pening.

The brasses should be of equal thickness from the face forming the joint to the back of the brass, so that the joint will be in the centre of the bore of the brasses. The respective faces forming the joint should be quite square with both the faces and sides of the brass, so that they will not spring the strap when they are keyed up, and so that, when the brasses are let together in consequence of the bore having worn, the faces may be kept square, and thus be known to fit properly together without having to put them together in the rod and on the journal to try them, which would entail a good deal of unnecessary labor.

To get the length of a connecting rod, place the piston in the centre of its stroke, and the distance from the centre of the crosshead pin to the centre of the crank shaft is the length of the rod from centre to centre of the brasses. Another method is to place the piston at one end of its stroke and the crank on its dead centre corresponding to the same end of the stroke, and the distance from the centre of the crosshead pin to the centre of the crank pin is the length of the rod.

To ascertain when the crank of a horizontal engine is upon its exact dead centre, strike upon the end face of the crank axle or engine shaft a circle true with the shaft, and of the same diameter as the crank pin : then place a spirit level so that one end rests on the crank pin and the other end is even with the outline of the circle; and when the spirit level stands true, the crank will be upon its dead centre.

The length of a connecting rod cannot be taken if the crank is placed in the position known as full power, because the position in which the piston would then be cannot practically be definitely ascertained; for the angle at which the connecting rod stands causes the piston to have moved more or less than half the length of the stroke when the crank has moved from a dead centre to full power, according to which end of the cylinder the piston moved from. If it was the end nearest to the crank, the piston moved less, if the other end, it moved more, than half of its stroke; so that in either case the piston stands nearer the crank than is the centre of the length of the cylinder when the crank is in the position referred to. This variation of piston movement to crank movement is greater in the case of short connecting rods than with long ones.

To fit a connecting rod to an engine, first rub some marking on the crank pin, and put the crank pin end of the rod on its place, with the brasses in and keyed properly up. The other end of the rod, being free, can be placed so as to touch against the crosshead pin, when the eye will detect if it will go into its place without any spring sideways; if it will do so, the rod may be taken off the crank pin, and the brasses, if necessary, fitted to the pin sufficiently to allow each to bear on the crown. But if the rod end will not fall into the crosshead journal without being sprung sideways, then move it clear of the crosshead, placing a side pressure on it in the direction in which it wants to go to come fair with the crosshead journal, and move it back and forth under such side pressure, which process will cause the crank pin to mark where the connecting rod brasses want filing and scraping to bring the rod true. The rod must then be taken off, and the brasses eased where the marking and the knowledge of which way the rod ought to go determine, the rod being placed on the crank pin as before, and the whole operation repeated until the rod "leads" true with the crosshead journal. The crosshead end of the rod must be fitted in like manner to the crosshead journal until the crank pin end of the rod leads true to the crank pin journal. The rod must then be put on its place, with both journals keyed up, and, if it can easily be accomplished, the engine moved backwards and forwards, the brasses being then taken out and bedded, when the rod will be fitted complete. A connecting rod which has both straps held by gibs and keys gets shorter from centre to centre of the bore of the brasses as it wears, and that to half of the amount of the wear. This is, however, generally rectified by lining up the brasses—that is, placing pieces of metal behind them (they may be fastened to the brasses if it is desirable)—which pieces are made of the required thickness to replace the amount of the wear of the brasses.

A connecting rod whose crosshead end has a strap with a gib and key, or, what is better, two gibs and a key, to hold it, the crank pin end having its strap held by bolts, and the key between the bolts and the brass, would maintain its original length, providing the wear on the crosshead brasses was as great as is the wear on the crank pin brasses; but since that on the latter is the greatest, the rod wears longer to half the amount of the difference of the wear between the crosshead and crank pin journals. If both the straps of a rod are held by bolts, the key of one end being between the brasses and the main body of the rod, and the key of the other end between the brasses and the crown of the strap, it would maintain its original length if the wear on both ends was equal; but this not being so, it wears longer, as above stated. When marking the length of the rod (that is, the circle on the brasses to set them by for boring), or when trammeling a rod to try its length, stand it on its edge; because if it rests on its broad face the rod will deflect, and appear to be shorter than it is; this is especially liable to occur in coupling or side rods, which are generally longer and slighter in body than connecting rods.

The oil hole of a strap for either a connecting or side rod should be in the exact centre of the space intended to be filled by the brasses. It will thus be central with the joint of the brasses, and from centre to centre of the oil holes, and will, therefore, represent the proper length of

Fig. 259.

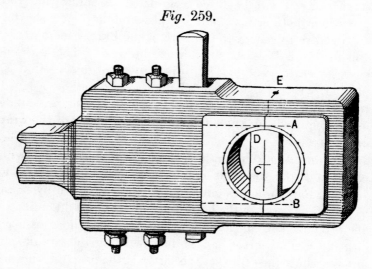

the rod. If the oil hole of the strap has been drilled to give the rod length as already explained, and new brasses have been fitted in, the bore may be marked out as in Fig. 259, there being a piece of wood or metal driven in the bore, and the line E being carried down by a try-square, and marked at D; lines A and B are obtained from the inside faces of the strap jaws, and from these the center is obtained, wherefrom to mark a circle to set the brasses by when chucking them to be bored.

In some cases, however, the brasses do not abut one

against the other, but are left open as in Fig. 260, and in this event the piece of wood must be placed across the bore as denoted by D in Fig. 260, the line B representing the center of the oil hole or the length of the rod.

The brasses should be so filed that lines as A in the

Fig. 260.

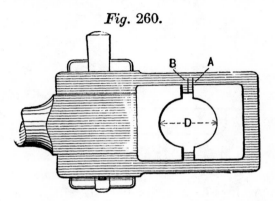

figure marked level with each face will be equidistant from B, and to accomplish this result, each brass must be laid upon a surface plate and tested with inside calipers, as in Fig. 261.

In letting brasses together to take up the wear, we must

Fig. 261.

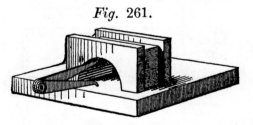

in brasses that abut against each other, or cone brass and brass, as it is termed, try them with calipers, applying the square both ways on the joint faces, for if the joint faces are at an angle instead of being square across, then driving up the key will spring the brass faces out of level one with another. Or if the faces are out of square instead of being square *across*, then driving home the key will spring the strap jaws open sideways.

In lining up brasses to set the key up in such a rod end

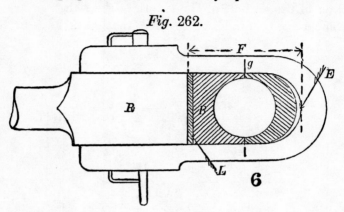

Fig. 262.

as is shown in Fig. 262, the liner L will be the one that determines the rod length, that at E simply serving to regulate the key height and not affecting the rod length.

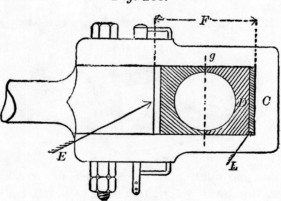

Fig. 263.

But when the strap is bolted to the rod end, as in Fig. 263, the back liner at L determines the rod length, and that at E is the one that raises the key.

DRIFTS.

Of drifts there are two kinds, one being a smooth round conical pin, employed by boiler makers to make the punched holes in boiler plates come fair, so that the rivets may enter, which may be aptly termed a stretching drift, and the other the toothed or cutting drift. Of the first it is to be observed that in some modern practice the holes in boiler plates are drilled, and are therefore more accurately spaced than it is practicable to punch them, thus greatly reducing the necessity to drift them; furthermore in the best practice the drilling is done after the plates have been bent into shape, thus dispensing altogether with the use of the drift-pin, which in the case of the badly matched holes found in punched plates greatly impairs the strength of the rivetted joint. The punching of a plate considerably weakens its strength at the narrowest section of metal, namely, between the hole and the edge of the plate, where the latter, being the weakest, gives way to the pressure of the punch. If one closely observes the surface of a piece of iron which is being punched, he will find that the scale on the surface of the iron round the hole, and especially between the hole and the edge of the plate, will be sensibly disturbed, showing a partial disintegration of the grain of the metal beneath, even if the punch is very sharp; but if the punch is dull, or the edge is in the least rounded by wear, the scale will fly off the surface of the metal in small particles, evidencing a considerable disturbance of the metal beneath and an equivalent weakening of the substance between the edge of the hole and the edge of the plate. If, then, after punching, the holes do not come fair, and the plain drift is employed to still further stretch the metal, not only is the weakening process greatly augmented, but the holes are stretched oval, so that the rivets do not completely fill them, however well the riveting may be performed. The use of the plain drift is therefore totally incompatible with first-class workmanship; hence a description of this tool will be altogether omitted.

Of cutting drifts, there are two kinds, the first being that shown in Fig. 264. A is the cutting edge, the width and thickness at C and B being reduced so that the sides of the drift may clear the sides of the hole. The drift is filed at A A, to suit the required hole, and tempered to a brown bordering upon a purple. The hole or keyway is then cut

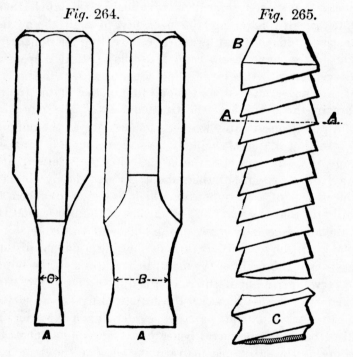

Fig. 264.   Fig. 265.

out roughly, to nearly the required size, and the drift is then driven through with a hand hammer, cutting a clean and true hole. Care must, however, be taken to have the work rest evenly upon a solid block of iron or (for delicate work) lead, and to strike the punch fair and evenly, otherwise a foul blow may break the drift across the section at C. This class of drift is adapted to small and short holes only, such as cotter ways in the ends of keys or bolts, for which purposes it is a very serviceable and strong tool. It must be freely supplied with oil when used upon wrought-iron or steel.

For deeper holes, or those requiring to be very straight, true, and smooth, the drift represented by Fig. 265 is used. The breadth and thickness of the section at A is made to suit the shape of the keyway or slot required. The whole body of the drift is first filed up parallel and smooth, to the required size and shape; the serrations forming the teeth are then filed in on all four sides, the object of cutting them diagonally being to preserve the strength of the cross section at A A. The teeth may be made finer, that is, closer together, for very fine work, their depth, however, being preserved so as to give room to the cuttings. To attain this object in drifts of large size, the teeth should be made as shown in Fig. 265, which will give room for the cuttings, and still leave the teeth sufficiently strong that they do not break. The head B of the drift is tapered off, so that, when it swells from being struck by the hammer, it will still pass through the hole, since this drift is intended to pass clear through the work.

The method of using this tool is as follows: The hole should be roughed out to very nearly the required size, leaving but a very little to be taken out by the drift, whose duty is, not to remove a mass of metal, but to cut a true and straight hole. To assist in roughing out the hole true, the drift may be driven lightly in once or twice, and then withdrawn, which will serve to mark where metal requires to be removed. When the hole is sufficiently near the size to admit of being drifted, the work should be bedded evenly upon a block of iron or lead, and oil supplied to both the hole and the drift; the latter is then driven in, care being exercised that the drift is kept upright in the hole. If, however, the hole is a long one, and the cuttings clog in the teeth, or the cut becomes too great, which may be detected by the drift making but little progress, or by the blow on the drift sounding solid, the drift may be driven out again, the cuttings removed, the surplus metal (if any there be in the hole) cut away, the hole and drift again freely oiled, and the drift inserted and driven in as before, the operation being continued until the drift passes entirely through the hole; for the drift will be sure to break if too much duty is placed upon it. After the drift has passed once through the hole, it should be turned a quarter revolution, and again driven through, and then twice more, so that each side of the drift will have contacted with each side of the hole (supposing it to be a square one), which is done to correct any variation in the size of the drift, and thus to cut the hole true.

The great desideratum in using these drifts is to drive them true, and to strike fair blows, otherwise they will break. While the drift is first used, it should be examined for straightness at almost every blow; and if it requires drawing to one side, it should be done by altering the direction in which the hammer travels, and not by tilting the hammer face (see Fig. 266).

Suppose A to be a piece of work and B and C to be drifts which have entered the keyways out of plumb, as shown by the dotted lines D and E. If, to right the drift C, it was struck by the hammer F, in the position shown and travelling in the direction denoted by G, the drift C would be almost sure to break; but if the drift B was struck by the hammer H, as shown, and travelling in the direction denoted by I, it would draw the drift B upright without breaking it; or in other words, the hammer face should always strike the head of the drift level and true with it, the drawing of the drift, if any is required, being done by the direction in which the hammer travels. When it is desired to cut a very smooth hole, two or more drifts should be used, each successive one being a trifle larger in diameter than its predecessor. Drifts slight in cross section, or slight in proportion to their lengths, should be tempered evenly all over to a purple blue, those of stout proportions being made of a deep brown bordering upon a bright purple. For cutting out long narrow holes, the drift has no equal, and for very true holes no substitute.

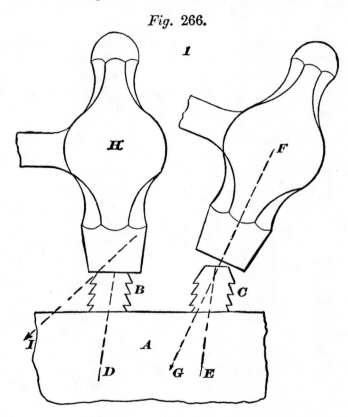

Fig. 266.

It must, however, be very carefully used, in consequence of its liability to break from a jarring blow.

REVERSE KEYS.

Crossheads, pistons, and other pieces of work which are keyed to their places upon taper rod ends, and are therefore apt to become locked very fast, are easily removed by means of reverse keys, which should always be employed for that purpose, because striking such work with a hammer, even supposing the work to be well supported underneath and copper interposed between the hammer and the work, is liable to bend and otherwise damage it with every heavy blow.

Reverse keys are simple pieces of steel, so shaped as to reverse the draft of a keyway, and are made male and female, as shown in Fig. 267, A representing the male, and B the female. The manner of using them is to insert them into the keyway, as shown in Fig. 268, in which A

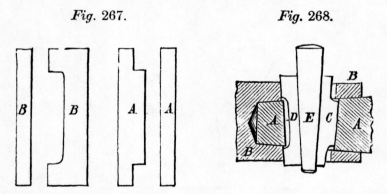

Fig. 267.   Fig. 268.

represents a taper rod end, B the socket into which A is fitted or keyed, C the male and D the female reverse key, and E an ordinary key. It will be found, on examination, that the insertion of C and D have exactly reversed the position of the draft of the keyway, so that the pressure due to driving in the key will be brought to bear upon the rod on the side on which the pressure was previously on the socket, and on the socket on the side on which the pressure was on the rod; so that driving in the key will key the socket out of instead of into its place.

The keyway in Fig. 268 is shown to have draft; that is, the proper key, when driven in, will bear one edge upon the edge of the keyway in the rod only, and not on the edge of the keyway in the socket at the small end of the cone; while at the large end, the natural key would bear against the edge of the keyway in the socket only. If however, this condition does not exist, and the edges of the key bear equally upon the cone and the socket (on both edges and all the way through), the keyway being a solid one, that is to say, having no draft, the reverse keys may be employed, providing that C is placed so as to bear upon the edge of the keyway on the large end of the cone only, and that D is placed to bear on the edge of the keyway at the small end of the cone on the socket only, thus producing a back draft, or clearance, as it may better be termed. The key E should be made long, and both it and the reverse keys should be made of steel and left soft.

SETTING LINE-SHAFTING IN LINE.

To set a length or line of shafting in line, first prepare a number of wooden frames or targets, such as in Fig.

Fig. 269.

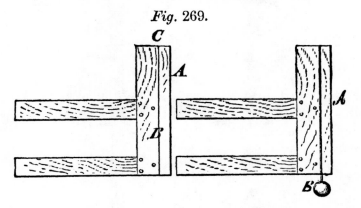

269, the outer edge A being planed straight, and there being marked a line B parallel to A. Upon this frame hangs the plumb-bob, shown at B, so that when the plumb line is fair with the marked line the edge A will stand vertical. Having erected these targets at each end of each length of shafting, we stretch a fine string or silk line beside the line of shafting, as in Fig. 270

Fig. 270.

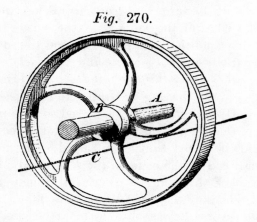

placing it about 6 inches below and on one side of the shafting, and adjusting it at first as nearly parallel to the

Fig. 271.

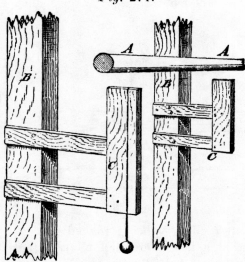

shaft as can be judged by the eye. If the line of shafting is of equal diameter at each end, we may set the stretched line equidistant from it at each end; while if one end is of larger diameter than the other, we set the line parallel to the shafting axis, and horizontally true as near as it can be set by a spirit level. The targets must now be adjusted as follows: the planed edge is brought up so as to just touch the stretched line, while its edge A is vertical, which will be known from the plumb line covering the line marked beneath it and parallel to edge A. Each target is set in this way and nailed fast or secured in any con-

Fig. 272.

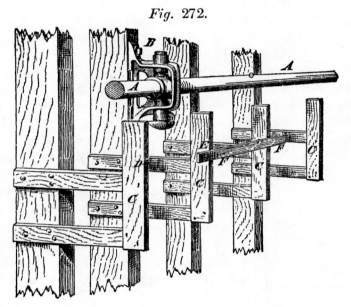

venient manner, as in Fig. 271. We have now in the planed edges A of the target a substitute for the stretched line, and forming a guide for the horizontal adjustment of the line of shafting. For the vertical adjustment we take a wooden straight-edge long enough to reach from one target to the next one, and beginning at one end of the shafting we place the flat side of the straight-edge against the planed edges of two of the targets at a distance of say 15 inches below the top of the shafting, and after levelling the straight-edge with a spirit level, we mark even with the edge of the straight-edge a line on the planed edge A of both of the targets. We then move the straight-edge so as to embrace the next target, set one end even with the line already marked on the second target, and set it true by a spirit level and mark a line on the edge of the third target, the straight-edge being shown in Fig. 272, in position to mark the line on the third target. By continuing this process we shall have marked a line across the edges of all the targets, and from this line the shafting may be set as follows: A square having its edges, A and B, at a right angle to one another has a line C marked upon it at a distance below the edge A of 15 inches (this being the distance we set the stretched line below the shafting axis), from line C on the square as a centre we mark below it a line F, in Fig. 273, distant from C to an amount equal to half the diameter of the line shaft, and if the shafting is parallel in diameter we may rub out line C and leave line F only on the square. All that remains to do is to apply the square to the edge of each target, and to the shaft, and when the line on the square coincides with the line on the target, the shafting is set true and level. For the horizontal adjustment all we have to do is to place a straight-edge on the edge of the target, as in Fig. 274, and adjust the shaft by a distance-piece D.

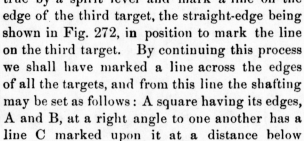

Fig. 273.

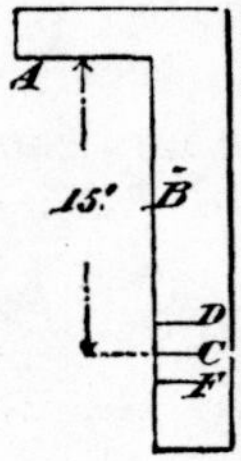

Fig. 274.

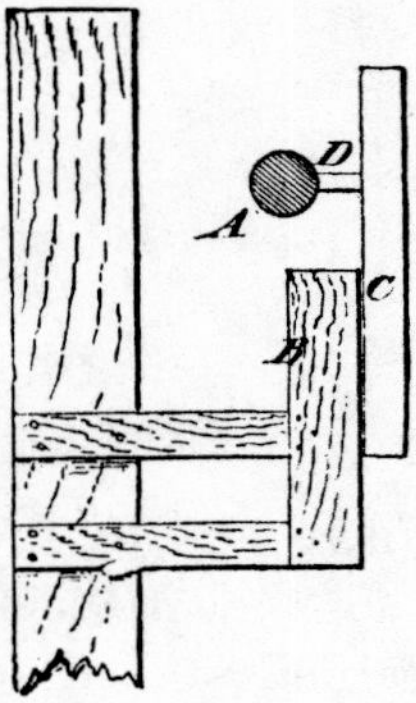

There are several points, however, during the latter part of the process at which consideration is required.

Thus, after the horizontal line, marked on the targets by the straight-edge and used for the vertical adjustment, has been struck on all the targets, the distance from the centre of the shafting to that line should be measured at each end of the shafting, and if it is found to be equal, we may proceed with the adjustment.

But if, on the other hand, it is not found to be equal we must determine whether it will be well to lift one end of the shaft and lower the other, or make the whole adjustment at one end by lifting or lowering it as the case may be.

In coming to this determination we must bear in mind what effect it will have on the various belts, in making them too long or too short, and when a decision is reached, we must mark the line C, in Fig. 273, on the gauge, accordingly, and not at the distance represented in our example by the 15 inches.

The method of adjustment thus pursued possesses the advantage that it shows how much the whole line of shafting is out of true before any adjustment is made, and that without entailing any great trouble in ascertaining it; so that in making the adjustment the operator acts intelligently and does not commence at one end utterly ignorant of where the adjustment is going to lead him to when he arrives at the other.

Then, again, it is a very correct method, nor does it make any difference if the shafting has sections of different diameters or not, for in that case we have but to measure the diameter of the shafting, and mark the adjusting line, represented in our example by C, in Fig. 273, accordingly, and when the adjustment is completed, the centre line of the whole length of the line of shafting will be true and level.

This is not necessarily the case if the diameter of the shafting varies and a spirit level is used directly upon the shafting itself. In further explanation, however, it may be well to illustrate the method of applying the gauge shown in Fig. 273, and the straight-edge C and gauge D, shown in Fig. 274, in cases where there are in the same line sections of shafting of different diameters.

Suppose then that the line of shafting in our example has a mid-section of $2\frac{1}{4}$ inches diameter, and is 2 inches at one, and $2\frac{1}{2}$ inches in diameter at the other end.

All we have to do is to mark on the gauge, shown in Fig. 273, two extra lines denoted in the Fig. by D and F.

If the line C was at the proper distance from A for the section of $2\frac{1}{4}$ inches in diameter, then the line D will be at the proper distance for the section of 2 inches, and E at the proper distance for the section of $2\frac{1}{2}$ inches diameter, the distance between C and D, and also between C and E, being $\frac{1}{8}$ inch; in other words, half the amount of the difference in diameters.

In like manner, for the horizontal adjustment, the gauge piece shown at D, in Fig. 274, would require, when measuring the $2\frac{1}{4}$ inches section, to be $\frac{1}{8}$ inch shorter than for the $2\frac{1}{4}$ inches section, while for the $2\frac{1}{2}$ inches section would require to be $\frac{1}{8}$ inch shorter than that used for the $2\frac{1}{4}$ inches section, the difference again being $\frac{1}{2}$ the amount of the variation in the respective diameters. Thus the whole process is simple, easy of accomplishment, and very accurate. If the line of shafting is suspended from the posts of a ceiling instead of from uprights, the method of procedure is the same, the forms of the targets being varied to suit the conditions. The process only requires that the faced edges of the targets shall all stand plumb and true with the stretched line. It will be noted that the plumb lines (shown on the target in Fig. 269 at B) are provided simply as guides, whereby to set the targets, and are put at about $\frac{3}{4}$ inch inside of the planed edge so as to be out of the way of the stretched line. It is of no consequence how long the stretched line is, since its sag does not in any manner disturb the correct adjustment; but in cases where it is a very long one it may be necessary to place pins that will prevent it from swaying by reason of air currents or from jarring.

The same system may be employed for setting the shafting hangers, the bores of the boxes being used instead of the shafting itself.

CHAPTER XVI.

MILLING-MACHINES AND MILLING-TOOLS.

THE position occupied by the milling-machine in modern practical mechanics is almost as important as that occupied by the lathe or planing-machine. In getting out work by the aid of either of the latter, the size and uniformity of the work depend upon the accuracy in measurement, and hence upon the skill of the operating artisan, hence a skilled and expert workman is necessary to the use of each lathe or planer. In the case, however, of a milling-

Fig. 275.

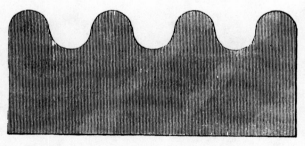

machine, the skilled mechanic has but to properly set the machine and the chucks necessary to hold the work, and a less skilful operator may be assigned to continue the operation of getting out any number of similar pieces of work, with the assurance that uniformity of size and form and equality of finish may be, with ordinary care, assured. Then, again, intricate forms and shapes of work may be exactly and easily duplicated by the employment of milling-tools, which would be impracticable were the same work operated upon by a planing-machine; especially is

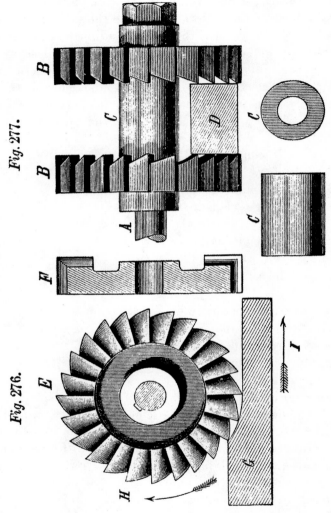

this the case in work of complicated form. Suppose, for instance, it were required to cut out a corrugated surface, such as shown in Fig. 275, it would be a difficult matter to produce, with a planing-machine, one such a piece of work quite true and with a smooth and polished surface, because the tool would be liable to spring from the broadness of cutting surface, which would, in the case of wrought-iron and steel, cause the tool to spring into the softer and away from the harder parts of the metal; and in the case of any metal it would be quite difficult to feed the tool so as to insure exactitude and avoid tool-marks at the junction of the cuts taken by the round-nosed and curved tools; whereas, with a milling-tool, properly made (and it is no difficult matter to make such a tool), the operation is so simple that it may be performed with comparatively unskilled labor.

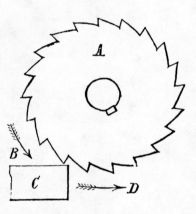

Fig. 278.

One of the main advantages of milling-tools is that the work will, in nearly all cases, be true, even, and smooth, even though the tool itself be a little out of true.

Suppose, for example, we require to mill the side faces of a rod, and we employ for the purpose the milling-bar and cutters shown in Fig. 277, in which A represents the spindle of a milling-machine, and B B are milling-cutters with the distance washer C interposed between them to regulate their distance apart; D representing a piece of work being fed between the revolving cutters B B. Now, it is evident that even were the cutters out of true, the pieces of work would all be cut to one size, because the projecting teeth of the cutters will come into contact with and operate upon each part of the surface of the work being operated upon, the only difference being that the work will be cut narrower with the same thickness or length of washer than it **would be were the washers true.**

In Fig. 276, E represents a view of the face of a milling-cutter, and F a sectional view of the same, while G represents a piece of work passing under the cutter and not between the cutters, as shown in the case of the work D. The arrow H denotes the direction in which the cutter E would require to revolve, and the arrow I the direction in which, in that case, the work would require to travel; from which it will be perceived that the lateral strain placed upon the work by the cut is in a direction to force the work back from the cutter, and this must always, in the use of milling-tools, be the case, and is a very important consideration for the following reasons:

From the breadth of cut taken by a milling-tool, and from the acute angle at which the teeth of the cutter strike the cut when the work passes below the circumference of the cutter, the strain due to the cut is immense; and were this strain in a direction to drag or draw the work below or towards the cutter, the latter would, from the spring of the spindle, rip into the work and tear its own teeth off. Thus, in Fig. 278, suppose A to be a milling-cutter revolving in the direction of the arrow B, and C to be a piece of work travelling in the direction of D, it will be readily perceived that there will be an enormous strain in a direction to force the work from its chuck or clamps and drag it under the cutter.

The work being held sufficiently firm, cannot, it is true, move in that direction faster than the rate of feed will permit; but the teeth grip the work, the cutter springs forward and attempts to ride like a spiked wheel over the work, and the cutter-teeth break from the undue pressure; and therefore it is that in milling work of every kind whatsoever, the direction in which the work is fed should be such as to tend to force the work away from the cut; or, in other words, the cutters should cut under the cut, not only because of the above imperative reasons, but for the following additional ones:

The skin of iron or brass castings and of iron or steel forgings is considerably harder than is the interior of the metal, in addition to which there is frequently scale in the one case and sand in the other to contend with, so that if the cutting edge of a tool comes into contact with the outer skin of the work, the keenness, and hence the cutting value of the tool or cutter, becomes rapidly impaired; and milling-cutters being expensive tools to make, it is desirable that their cutting edges and qualifications be preserved as long as possible. Suppose, therefore, that in Fig. 279, from A to B represents the depth of cut on two

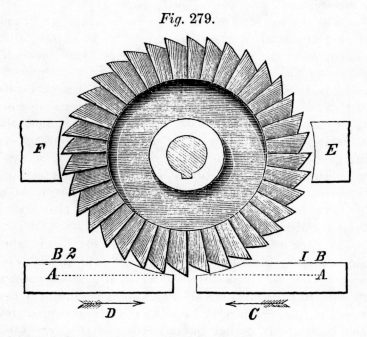

Fig. 279.

pieces of work, one travelling beneath the cutter in the direction of the arrow C, and the other in the direction of the arrow D, and that the upper surfaces B, in each case, have a hard surface-skin upon them: it becomes apparent then that in the case of the piece represented at 1, the cutter-teeth will, after the cut has once started, meet the soft metal and cut under the skin till the cut has ended, so that, save at the very commencement of the cut, the cutter-teeth would never meet or come into contact with the hard surface-skin; while in the case of the piece of work 2 the teeth would in every instance strike the hard skin first. If the piece of work E were held in the position shown, it would strike the scale, whichever way the cutters ran or the work was fed; and the same remark applies to the piece of work F. There is this difference, however, between the two latter positions: with the cutter revolving in the direction shown, the strain of the cut would be in a direction to lift E from the machine-table, rendering it very liable to spring and difficult to cut; while the strain on F would tend to force it down upon the table, which would be far preferable.

When the side faces of the cutters operate, they must be made right and left—that is to say, the teeth of one cutter must slope in the opposite direction to those on the other cutter, so that when the two are placed opposite to one another, as shown in Fig. 277, the teeth of both will stand in a direction to accommodate the direction in which the cutters revolve. To cut side faces of any required width, we have only to vary the width apart of the cutters by the washer C, in Fig. 277; while, to cut curves and shoulders, the periphery only of the cutters can be used. Thus, suppose it were required to cut out the form shown in Fig. 280, the outline of the cutter would require to be as shown in Fig. 281, but it would be a tedious and difficult matter to get up a solid cutter of such a shape on account of the difficulty of cutting the teeth; hence, all such compound forms are produced by making separate cutters, each of its requisite form, size and width, and then placing them together to make up the whole. Thus the figures from 1 to 8 each represent a separate cutter. It is obvious then that there is scarcely a limit to the forms capable of being smoothly cut and uniformly reproduced by such cutters. The Morse Twist Drill Company cut the threads upon their taps, and give the sides of threads a slight amount of clearance back from the cutting edges by the use of milling-tools, producing a tap equal in every respect to those producible in the lathe, and being remarkable for uniformity of size and finish.

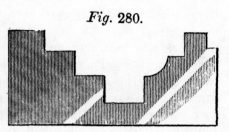

Fig. 280.

Milling-cutters of small size are made of solid cast-steel; for larger sizes, the body is made of wrought-iron, while the faces whereon the cutting-teeth are to be formed have steel welded on them.

After the cutters are bored and turned to the requisite size and shape, the spaces necessary to the formation of the teeth may be cut by a milling-cutter; and here it may be well to note that it is advisable to keep the teeth sufficiently wide apart to give plenty of room for the cuttings

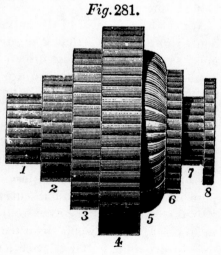

Fig. 281.

to escape; even in cutters for gear-wheels, coarse teeth— that is, those wide apart—will cut quicker and smoother than fine ones, and have the advantage that they entail much less labor in both the manufacture when new and resharpening when dull. After the spaces are cut out and the teeth formed, the cutter must be carefully hardened and tempered to a straw color.

Very long toothed cutters, such as reamers, are apt to warp in the hardening, getting out of round as well as out of straight. They may of course be made true again, by the ordinary grinding process, or the following plan may be adopted to straighten them previous to sharpening them in the usual manner:

The reamer, after being hardened, should be revolved rapidly in the lathe in one direction, while an emery-wheel revolves at a high speed in the opposite direction, as shown in Fig. 282, A representing the reamer and B the emery-wheel; the emery-wheel should be fed to the reamer just sufficient to true the cutting edges. For ordinary cutters clearance and sharpening may be given to the teeth as follows: Beneath a revolving emery-wheel, and quite parallel and true with the spindle on which the emery-wheel revolves, there is provided a stationary adjustable mandril of such a size as to neatly fit the centre hole in the cutter to be operated upon; which mandril is of a sufficient length to permit the cutter to slide along it and stand wholly on either side of the emery-wheel. The height of the mandril is adjusted so that the emery-wheel will, when brought into contact with the cutter (while the latter is upon the mandril), take off just sufficient to sharpen and give clearance to the teeth. Some guide is necessary to insure that the teeth of the cutter shall pass under the emery-wheel in an exactly uniform position, which is accomplished by providing an adjustable stationary guide or gauge, against which the radial or front face of the tooth is held while it is being

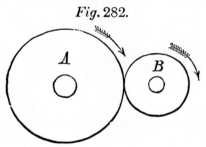

Fig. 282.

ground. The operation is thus to place the cutter on the mandril and adjust the latter to the requisite height, to then adjust the guide so that when the cutter is moved forward it will first come into contact with the guide, against which the cutter is held by the hand, while it is at the same time passed under the emery-wheel. It is obvious that by this means either circumferential or side face teeth may be sharpened and maintained true if the bearing of the cutter upon the mandril is sufficiently long and of sufficiently accurate fit to keep the cutter steady. There are other devices for taper cutters in which the latter are stationary and the emery-wheel traverses along the teeth, which plan is for taper cutters preferable to that first described; the principles involved are, however, the same in both cases.

It must be remembered that, in using the emery-wheel for this purpose, it *must* run under its cut for the reasons already explained by Fig. 278 and its accompanying explanation.

It is obvious that in the case of long cutters having circumferential teeth, the excessive strain due to each tooth striking the cut will cause the mandril carrying the work to spring away from the cut, the effect being that the finished surface of the work will be slightly waved. To remedy this defect, the teeth of the cutter should be made to run slightly spiralled, and not straight across the length of the cutter, so that the cutting edges will be taking and leaving the work continuously, and hence the spring above referred to will be at all times equal. The same object is obtained in compound cutters, such as shown in Fig. 281, by cutting the key or feather-ways in the cutters, so that their teeth will not stand in a line one with the other.

CHAPTER XVII.

GRINDSTONE AND TOOL GRINDING.

GRINDSTONES are employed for three purposes: to smooth surfaces, to reduce metal to a given thickness, and to sharpen edge tools.

The following are the various kinds of grindstones, and the uses for which they are best suited:

Ohio, Nova Scotia and English stones are those principally used, and each of these are cut to various sizes, and have different degrees of coarseness and fineness of grit. *New Castle* stones have a yellow color and sharp grit, the fine soft ones for grinding saws to gauge thickness, and the coarser for rougher work, as grinding sad-irons and springs, for face stones in nail works, and for castings (dry grinding). *Wickersly*, grayish-yellow color, for saws and cutlers' work generally, having a very soft grit and hence not liable to heat the work and draw its temper. *Liverpool* (or Melling), of a red color and very sharp grit. for edge tools generally. *Nova Scotia*, blue or yellowish gray color, being of all grits from the finest and hardest to the coarsest and softest, used for general grinding as well as for machinist tool grinding. *Bay Chaleur* (N. B.), of a uniform blue color and soft sharp grit, for cutlery and for fine edge tools. *Independence*, grayish white color, soft loose grit, for edge tools requiring a fine edge, and for the dry grinding of castings. *Huron* (Michigan), of a uniform blue color and fine sharp grit, best suited for tools requiring a fine sharp edge as carpenters' tools. It is desirable that the fineness or coarseness of the grit be uniform throughout the stone, because a hard spot will wear the least and become a projection, while a soft one will wear the quickest and become a depression, either of which is a defect preventing the production of true and smooth grinding.

A stone having a coarse open grit will cut the most freely and remain the most true, but will not last so long.

Stones having a close though coarse and hard grit are apt to become coated with particles of the metal cut from the work, which is remedied by grinding a round iron bar of small diameter, the small area of contact acting to dislodge the particles and clean the stone.

Grindstones that are used wet (as is generally the case) should be kept supplied with water when in use, but not allowed to stand in water when at rest, because the part immersed will absorb water and become the hardest, which will cause it to wear the least, thus throwing the stone out of true.

Furthermore, if the stone be overrun at a quick speed the side that is water-soaked becomes the heaviest, and this may throw the stone sufficiently out of balance to cause the unequal centrifugal force generated at a high velocity to burst the stone.

The surfaces of stones used for sharpening machinists' tools should be kept as smooth as possible, while those used for removing metal or taking the skin from metal or similar work, where the object is to remove the metal as quickly as possible, are what is termed *hacked;* that is, they have indentations cut in them with a tool similar to a carpenter's adze. This hacking is usually performed most on the high parts of the stone, so as to cause them to wear the most and by this means keep the stone true. This class of stone is usually of the harder and coarser kinds of grit, the diameter of the stone ranging from 5 to 6 feet, and the face from 7 to 12 inches wide, and are run at speeds sometimes as great as 2000 feet per minute.

A grindstone should run as true as it is practicable to have it. This is attained by either fitting to the stone truing devices, or truing the stone (at intervals) by hand Sometimes two stones are so set as to enable their perimeter to touch, the speeds of rotation varying (or the direction of rotation may vary); hence one stone keeps the other true. In this case the best results are obtained when one stone only is used, the other (which may be a small one) being a truing stone, which is preferably caused to traverse laterally back and forth across the other stone so as to keep the face of both stones at a right angle to their planes of rotation.

In some cases grindstones are fitted with traversing rests which hold the work and traverse it back and forth between two grindstones. One of the two stones is speeded up to about 2000 feet per minute to perform the grinding, and the other (usually of smaller diameter) is driven by a gear wheel so as to hold the work from rotating or cause it to rotate at a slow speed, as about 80 feet per minute. The smaller stone is so carried that it may be traversed to or from the larger one, to suit different thicknesses of work and to put on the cut.

For very accurate grinding, as for grinding the surfaces of engravers' plates, the grindstone is, in the best practice, mounted upon a machine somewhat similar to a planer—the rotating stone occupying the position of the planer tool. The work, however, simply lies upon the table and is traversed back and forth beneath the stone by hand, which obviates the spring that would ensue from the pressure of work-holding devices.

A grindstone, for tool grinding, having no truing device, may be turned up with a piece of hardened steel or a piece of gas or other iron pipe about $\frac{3}{4}$ inch in diameter. A piece of iron is laid across the grindstone trough to form a rest, and the tool or pipe is held at a considerable angle to the stone, the cutting end being well below the surface of the trough. By this means the pipe is not liable to take too deep a cut or to be wrenched from the hands, as it will give way somewhat to the projections on the stone. The tool should be about 4 to 5 feet long, and should be fed to its cut by being slowly revolved, which will not only feed it but also keep the end of the tool square as is necessary. It should be pressed firmly down to the fulcrum bar or rest to prevent it from slipping. After the stone is turned true and to shape it should be smoothed by the edge of a straight thin piece of sheet-

Fig. 283.

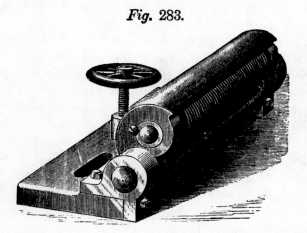

iron, which should be moved laterally, to prevent the edge from grinding, to fit the projecting rings on the stone.

During this turning operation the stone should be quite dry and should revolve at a slow speed, and, to prevent the pipe from becoming red hot, it should be occasionally dipped in water.

The intermittent truing of a stone is, however, objectionable, because for the proper sharpening of tools the stone should be kept continuously true, and for this purpose we have the device shown in Fig. 283. It consists of a frame carrying a threaded hardened steel roll bolted to the trough, as in Fig. 284.

GRINDSTONE AND TOOL GRINDING.

The piece carrying the roll is pivoted below the roll, hence by turning the screw (by means of the hand wheel shown) the roll is brought into contact with the stone surface, the thread crushing the projections on the stone and thus keeping it true. The friction between the surfaces causes the threaded roll to revolve and prevents its rapid wear. The advantage of this device is that the stone may be kept true while in constant use, and, as the device works with water, the dust and dirt due to turning

Fig. 284.

with a tool is avoided; furthermore, the stone is maintained true across its width, which is highly advantageous in grinding machinists' tools, especially those for coarse feeds where it is difficult to grind the nose of the tool flat and straight. Since the device can be applied to or disengaged from the grindstone surface at pleasure it does not unduly wear it away, and indeed causes, if properly used, no more wear than is essential to keep the stone true. The device should be placed upon that side on which the stone leaves the trough when rotating.

If a stone does not run true the tool will dip into the depressions and rise over the projections, rendering the operation unsteady; rounding the tool facet, and destroying the flatness that is necessary to the production of a keen cutting edge.

The face of a grindstone for flat surfaces may be flat across, or slightly rounding, as in Fig. 285, but in no case hollow, as in Fig. 286, for reasons which will be explained presently.

If in grinding a tool it is held in such a position that the circumference of the stone runs towards the operator, the grinding can be performed quicker and, as a rule,

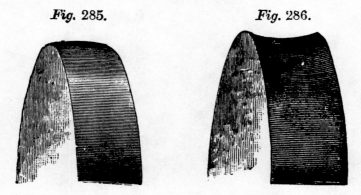

Fig. 285.   Fig. 286.

better; but it is, in many cases, quite dangerous, because the edge of the tool is liable to catch in any soft places or spots in the stone and to be dragged from the operator's fingers; sometimes it will force them down with violence to the tool-rest, rendering them liable to injury by being caught between the rest and the stone. The rest should be bolted firmly to the trough, and should stand on such side of the stone that the stone runs towards the top of the rest, this top also being above the level of the centre of the stone.

In determining upon which side of the stone any given tool should be ground, the workman takes into considera-

GRINDSTONE AND TOOL GRINDING.

tion the following: the shape of the tool, the amount of metal requiring to be ground off, and the condition of the grindstone.

Upon the edge of a tool which last receives the action of the stone, there is always formed what is termed a feather edge, that is to say, the metal at the edge does not separate from the body of the metal, but clings thereto in the form of a fine ragged web, as shown in Fig. 287, in which A represents a grindstone running in the direction

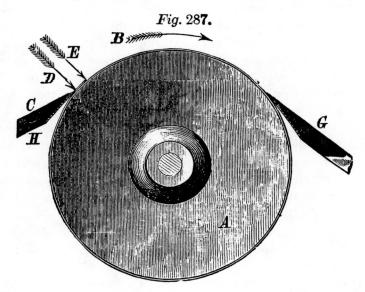

Fig. 287.

of the arrow, B, and C represents a tool. If now we take a point on the circumference of the stone, as say at F, it should leave contact with the tool at the point denoted by D; instead of doing this, however, the metal at the extreme edge of the tool gives way to the pressure and does not grind off, but clings to the tool, leaving a web, as shown from D to E; whereas, if the same tool were held in the position shown at G, the stone would meet the tool at the edge first, and would cut the metal clear away and not leave a feather edge. Now the amount of the feather edge will be greater as the facets forming the edge stand at a more acute angle one to the other, so that, were the facets at a right angle (instead of forming an acute wedge, as shown in Fig. 287) the feather edge would be very short indeed. But in all cases the feather edge is greater upon soft than upon hard metal, and is also greater in proportion as the tool is pressed more firmly to the stone; hence the workman conforms the amount of the pressure to suit the requirements by making it the greatest during the early grinding stage when the object is to grind away the surplus metal, and the least during the later part of the process, when finishing the cutting edge, and hence he obtains a sharper tool, because whatever feather edge there may be breaks off so soon as the tool is placed under cutting duty, leaving a flat place along the edge. It would seem, then, that faces which can be ground in the position relative to the stone, shown in Fig. 287 (that is, with the length of the cutting edge lying across the stone), and being upon a tool of shape similar to that shown in the figure, should always be ground with the stone running toward the cutting edge, as shown in the figure at the position denoted by G. This is the case, providing that the stone runs very true and contains no soft or hard spots of sufficient prominence to cause the cutting edge to catch, which would, as already stated, render the operation dangerous. These unfavorable conditions, however, are always more or less existent, under average conditions and to such an extent as to forbid the holding of the tool to the stone with the amount of pressure necessary to remove such a quantity of metal, as is necessary in the earlier stages of the grinding operation. Furthermore, if the edge of the tool does catch in the stone, the damage to that edge, by being ground away, is very serious and entails a great deal of extra grinding to repair it, and at the same time incurs a rapid using-up of the tool. Another consideration is that it is much easier to hold the tool steady, under ordinary circumstances, in the position shown at H, than in that shown at G; and with a bad stone it is altogether impracticable to hold it as at G. Here, however, another consideration occurs, in that the surface of a grindstone is rarely level across the width of the perimeter of the stone, unless the stone has a truing device attached to the frame, which at present is very largely the exception. As a rule the face of the stone is made rounding in its width because there is the most wear in the middle, and it is very undesirable to have the stone hollow across. Suppose, for example, that we have a stone that is hollow, as in Fig. 284, and one that is rounding across the perimeter, as in Fig. 285; then to grind such a tool as is shown in Fig. 287, as say a plane blade, we may move it slowly across the width of the stone, and the highest part of the stone will act upon all parts in the width of the blade; but we cannot, by any method, grind such a tool upon the hollow stone without leaving the cutting edge rounding in its length, or else leaving the ground facet rounding in its width or depth, the latter being the case when the cutting edge is held in line with the plane of rotation of the stone.

In grinding pointed tools, as centre-punches, scribers, etc., grooves are very apt to be worn in the stone unless the tool be moved back and forth across the stone face, or held at an angle to the plane of rotation.

The reason, when truing up a stone by hand, for leaving it rounding across its face is that the middle is more used than the edges, and the wear is, therefore, correspondingly greater in the middle. This causes the stone to gradually wear first flat across and then hollow, necessitating that it be turned up though it may run true.

When a stone is uneven across its width, the operator, no matter which side of the stone he is using, holds the length of the cutting edge of the tool at an angle to the width of the stone, as shown in Fig. 288, placing the tool in the most level part of the grindstone surface. By doing this he effects two objects: first, he obtains a level spot upon the stone more readily, and secondly, he diminishes the formation of a feather edge. The first is because it is obvious that, in removing a given amount of metal, there will be more abrasion upon the stone in proportion as the operating area of the stone is diminished, hence the work-

Fig. 288.

man selects the highest part of the stone whereon he can find a suitable surface, and by moving the tool across the face wears down the asperities (while he is roughing out the tool) so as to obtain as smooth a surface as possible for finishing process. If he held the tool still instead of giving it lateral motion, it would grind away in undulations or grooves conforming themselves to those on the abrading surface of the stone, and the roughing process would have but very little effect towards leveling the stone. Referring now to the second advantage named,

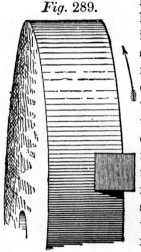

Fig. 289.

it will be readily observed that, if he held the length of the cutting edge in a line with the revolutions of the stone, as in Fig. 289, there would be no tendency to leave a feather edge, except at the corner of the edge where the stone leaves contact with the tool, and this would be of little or no consequence. The question naturally arises, then, why not grind the tool in that position, which would require a very small flat or smooth space in the width of the stone and would avoid the formation of a feather edge. The answer to this is that it would be very difficult to grind the surface of the tool level, as will be perceived from the side view of the operation as shown in Fig. 290, in which A represents

Fig. 290.

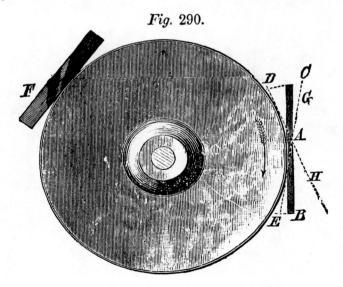

the tool enlarged so as to make the engraving clear. To bring the whole length of the cutting edge to bear upon the stone it is necessary to move the tool from C to D, and from B to E, as denoted by the dotted arcs at D, E; and if during this movement the tool remains an

Fig. 291.

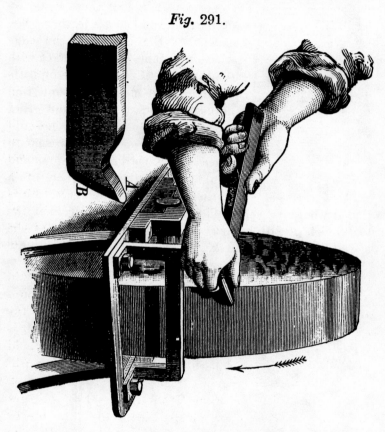

instant too long in either of the positions indicated by the dotted lines, G, H, or at any time during the motion, **a** hollow spot will be ground upon the tool at the point of contact between the stone and the tool; furthermore the grinding operation is not very accessible to the eye and hence any irregularities are not very easily corrected.

For these reasons it is impracticable to grind in this position any cutting edge requiring to be a straight line and having sufficient length to render much motion in the direction of D, E, a necessity. Furthermore it is very difficult to hold a tool steadily in position shown in Fig. 290, and as a consequence no satisfactory result can be attained unless by the aid of a device whereon to rest the hand; such a device is called a rest and is shown in Fig. 291. Now suppose we have a tool of the form shown at C in Fig. 291, requiring to be ground on the faces, A and B; then it is evident that A can only be ground with the body of the steel, C, out of the way of the body of the stone, and hence in the position shown in the figure, in which position the tool may be held and pressed firmly to the stone. It is necessary, however, to rest the hand upon the rest and hold the tool exactly in the position shown, so that if the tool catches in the stone and is forced from the hand it will not carry the fingers with it, and wound them by jamming either against the stone or the rest, or force them between the two. It would seem advisable to rest the tool upon the rest without the intervention of the hand, but such is not the case, because the operator would not have sufficient control over the tool and it would almost assuredly catch in the stone. By interposing the hand between the tool and the rest, the sense of feeling is brought into play, guiding the operator just how to hold the tool to prevent its catching in the stone and admonishing him when the conditions possess any elements of danger, which become instantly known from any difficulty in holding the tool steady against the grip of the stone or from a disposition of the upper edge of the tool, which the stone meets, to turn in towards the stone.

CHAPTER XVIII.

LINING OR MARKING OUT WORK.

When work is got out by means of special machines, or in special jigs, chucks, or appliances, it is generally unnecessary to denote its shape or dimensions by lines. But in large work, such as marine engine work, there are rarely a sufficient number made of precisely one pattern to make it pay to get these special machine tools or appliances, hence the work requires to be marked out by lines.

So likewise in the case of repairs for all save the small class of machines, such as sewing machines, the work requires to be lined or marked out because the original dimensions must be varied to accommodate the wear of the parts.

In the general machine shop the lining out of work forms an important part of the manipulation. In the case of a very simple piece, such as say a square bar, the lining may be dispensed with because the lineal measuring rule will demonstrate whether the work is large enough to permit of being cut to the required dimensions. But in irregular shaped bodies it is necessary to mark out the work by lines which serve to set the work true on the machine tools, and to denote the dimensions to which the work should be cut to reduce it to its required form.

In giving examples of the processes of lining out work, it has been thought best to let them be of the various parts of an engine, and with this view each part of a simple engine that could be utilized to represent a certain class of work has been selected.

First, however, let it be noted that while the principles applied in marking or lining out are the same as those involved in mechanical drawing, yet the application is entirely different. The draughtsman may obtain his centres and lines for one view or side of a piece by projecting those from another view or side of the work, whereas in marking out the work the lines must be transferred from one side of it to the other; or, in many cases, the lines on one side may be entirely different from those on the other, and yet require to be definitely located with reference to the same.

The lining out of work requires also to be more accurately performed than do the lines on a drawing, because the variation to an amount of the thickness of a line may involve the spoiling of a piece of work or entail a great deal of extra labor in fitting the parts together. Suppose, for example, a block and a strap requiring to be fitted together are marked by lines: the block being marked the thickness of the line too large, and the strap the thickness of the line too narrow. When the work comes to be fitted there will be the thickness of the two lines to file away to make the strap fit to the block. Furthermore, unless lining out or marking out work by lines be very accurately performed, an element of uncertainty is engendered, and the machine operators, instead of cutting away the surface metal to split the line, will leave the lines in as evidence that they have not removed too much metal, and as a result the filling operations are again increased.

A marker-out, as the operative is termed, should not only be one capable of great exactitude in his measurements, but should also be an expert workman at the lathe, vise, planing machine, and drilling machine; because it is by his lines that the work is chucked, and hence he should know the very best method of chucking or holding the work in each of the machines. Furthermore a line over and above those necessary to define the outline of the work is often necessary for use as an assistance and guide in chucking it. Upon the truth of this lining, in many cases, will the truth of the finished work depend, and even in those instances where the method of chucking will correct any inaccuracy in the marking-out, the usefulness of the latter is almost entirely destroyed, because the lines will become entirely removed on one side, and left fully in on the other side of the work. If, however, the marking-out is performed reasonably true, one of its main elements of usefulness consists in that it denotes if there is sufficient excess of metal upon the piece of work to permit of its being cleaned up all over. But if there is any one part of the work scant of metal, as is sometimes the case in forgings of unusual and irregular form, the marking-out requires to be very true, and may be made to just save a piece of work that otherwise would have been spoiled. By accommodating the marking to some spot or place in the work, which will only come up to the full size by throwing the whole of the rest of the lines towards the opposite side of the work, a costly piece of forging may be saved from the scrap heap. And again, in castings where the surface appears spongy, showing the presence of air holes beneath the surface, or in forgings where the surface may indicate that a weld is not perfect upon one side, the whole of the marking-out should be performed with a view to take off as much metal as possible on the faulty side. In other work there may be a part very difficult to turn or plane on account of the conformation of the job; in which case the marker-out, foreseeing such to be the case, will so place the lines as to give as little to come off that particular place as possible, disregarding the excessive heavy cut or amount of metal which it may be necessary to cut off other and more accessible parts of the work. There are many other considerations, which need not be here enumerated, all tending to show that a

LINING OR MARKING OUT WORK.

marker-out should be a master hand at the various branches of his business, and possess much judgment and experience.

TO MARK AN ELLIPSE.

Draw the line A B, Fig. 292, equal to the required length of the ellipse. Bisect it by the line C D, which must stand at a right angle to it, be equal in length to the required width of the ellipse, and extend an equal distance on each side of it. With a radius of one-half the required length of the ellipse mark from C (or D) as a centre the arc F H G, and at the points of intersection of this arc with the line A B (that is at F and G) drive in two pins. Drive in a pin at C and pass a piece of fine

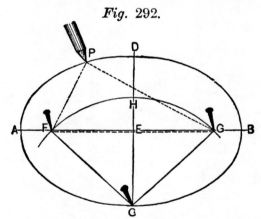

Fig. 292.

twine around the pins at A F, G, and C, and tie it tight enough to prevent any slackness.

Remove the pin at C (which is only employed for convenience in tying the twine to its proper length without being under tension or having any slack). Take a pencil: move it outwards from the pins F G until the twine is drawn straight, then sweep it around and its point will describe an ellipse. In the Fig. the pencil is shown at P, the position of the twine when the pencil is at that point being denoted by the dotted lines.

The pencil must be held vertical while tracing, otherwise the figure will not be true. To assist this a piece of wood may be laid and slid on the surface, on which the ellipse is traced, and the pencil held against the piece of wood.

TO FIND POINTS THROUGH WHICH THE CURVE OF AN ELLIPSE MAY BE DRAWN.

Let A B, C D, Fig. 293, be the respective diameters of the curve: mark the parallelogram L M N O meeting the ends of the axes as at A B, C D. Divide A L, A N, S M, and B O into any number of equal parts, and number them as in the Figure. Divide A S and D S into four equal parts: numbering them from the ends towards the centre. From the points of division in A L and B M draw lines to the point C, and from the points of division in B O and A N draw lines to the point D.

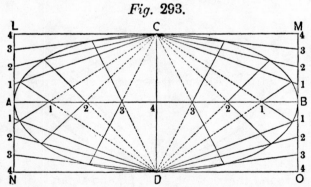

Fig. 293.

From the point D draw lines passing through the points 1, 2, 3 in A S to intersect the lines 1, 2, 3 drawn from the points of division on A L; also from point D through D S to intersect the lines from B M. These points of intersection are points in the curve.

From C, through the divisions 1, 2, 3 on S A and S B, draw lines to intersect lines 1, 2, 3 drawn from A N and B O: the points of intersection being points through which the other half of the curve may be drawn.

LINING OR MARKING OUT WORK.

The tools employed by the marker-out are as follows:

For plugging holes so that compass points may be supported within the hole, disks of lead such as are shown in Figures 294 and 295 are employed. They may be stretched larger or compressed smaller in diameter to suit any required size of hole, by a few blows with the hand-hammer, and the lead will conform itself to the uneven shape of the hole, and will therefore hold fast and not be liable to move; and, furthermore, a few such blows will deface any lines which may have been made upon the face of the lead in service upon a previous piece of work. Again it may be necessary to first mark a centre line, and subsequently other lines; and then drawing a wet finger across the old lines on the lead will dull them, while the newly made ones will be bright, and thus remain distinct. For holes that have been trued out, similarly shaped pieces of sheet brass may be used, the form shown in Fig. 295 being for the larger, and that shown in Fig.

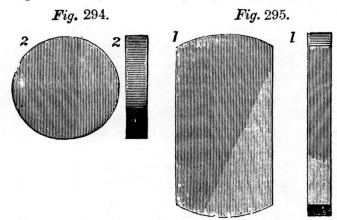

Fig. 294.   *Fig.* 295.

294 for the smaller sized holes; these brass pieces may be filled up very true, and have a centrepunch mark in their exact centre, thus obviating the necessity of finding the centre at each time of using.

For use on holes of comparatively large dimensions,

that is to say, above 4 inches in diameter, the centre piece shown in Fig. 296 is very convenient. A represents a piece of wood, and B, a small piece of tin or sheet iron, having its corners bent up so that they may be driven into the wood and thus made fast in position to receive the

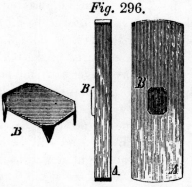

Fig. 296.   Fig. 297.

centre. Such a centre is very easily and readily made, and may be used on rough or finished work. If the surface of the work upon which either of these centres is used is flat, the ends of the centres must of course be also flat; and in the case of the last described, a piece of paper, leather, or other material may be inserted in one end to make up any small deficiency in the size. The centrepunch used for marking-out should be as shown in Fig. 297, the object of making its diameter so small toward the point being that it shall not obstruct a clear view of the line. A heavier centrepunch may of course be employed to increase the size of the centrepunch marks when the same is necessary. The hammer should also be a small one, weighing about one-quarter of a pound, with a ball face to efface any centrepunch marks erroneously marked or to be dispensed with, an ordinary hammer being employed to perform any necessary operation other than the simple marking-out.

TO DIVIDE A STRAIGHT LINE INTO TWO EQUAL PARTS.

From each end of the line we describe arcs of circles, as at F in Fig. 298, adjusting the compasses so that the two arcs meet at the given line, as shown in the figure.

Fig. 298.

By then resting the compass points at the coincidence of the arcs F, and describing the arcs B C, the latter will cut the ends of the line, as shown in the figure.

TO DIVIDE A STRAIGHT LINE INTO A NUMBER OF EQUIDISTANT POINTS.

Let A B, in Fig. 299, represent a line to be divided into 10 equal divisions. With a pair of compasses set

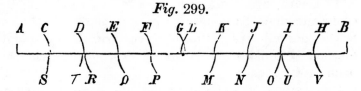

Fig. 299.

as near as may be to $\frac{1}{10}$ the length of the line measured upon a finely divided rule, and starting from the end A of the line step off on the line, and mark above the line the arcs C D E F G. From B step off and mark the arcs H I J K L, and if the compasses are correctly set, G and L will join at their point of coincidence with the line. If not, mark a point as shown in the figure, upon the line and midway between G and L. Since G and L overlap each other the compass points are too far apart, hence they must be corrected, which may best be done when the error is a very fine one by oilstoning them on the outside. With the corrected compasses and from the dot as a centre, step off upon the line divisions P O R. From the end A of the line step off divisions S T, and midway between S and T upon the line mark another dot, and these two dots will be correct points of division, notwithstanding that the compass points are shown in both cases to be incorrectly set. The want of coincidence of T R, which do not meet at their point of coincidence with the line, shows the compass points to be too near together, hence in this case those points must be corrected by oilstoning them on the inside. This being done from the dot at G L as a centre, step off on the line the divisions in M N O and mark the arcs, then from the end B of the line mark off V U, and midway between O U is another point of division. The error shown to exist in the compass point being again corrected we may from these four points mark off the intermediate ones. By this method the division may be proceeded with while correcting the compass points. If, however, the number of divisions is an odd one instead of an even one as in this example, the compasses must be stepped from each end of the line as before, and adjusted until the space forming the middle division is equal to the distance at which the compass points are set.

But suppose it be required that the points of division vary in regular order as in the case of a piece requiring say 60 holes in a given length, the distance between two given holes being 1.57 inches, and the two next 1 inch, and so on continuously.

The total number of holes must in this case be an even one; hence we mark off, by the rules already given, one half of that total number, making them equidistant all round the circle or circumference, as the case may be, which points will represent the distance apart of the holes that are widest apart, as the holes, A, B, C, D, and E, in Fig. 300, amounting in our example to 30 in number. We then set our compasses to the required distance apart of the two holes nearest together; and commencing at A in Fig. 300, we mark the centre for the hole, F, and from the centre of the hole, B, the centre of the hole, G, and

LINING OR MARKING OUT WORK. 369

so on, continuing all round the circle, but taking care to mark the new centre in each case in advance of or behind the points, A, B, C, etc., according to the manner in which

Fig. 300.

the first of the holes nearest together was marked. Thus in Fig. 300, the points, F, G, H, and I, are marked to the right, in each case, of the points from which they were struck.

Before proceeding to mark out a piece of work, it should be roughly measured so as to ascertain, before having any work done to it, that it will clean up. The square should also be applied to see if it is out of square, and thus to find out if it is necessary to accommodate the marking out to any particular part that may be scant of material (or stock, as it is often termed). The surface of the work should also be examined; so that, if any part of it is defective, the marking off can be performed with a view to remedying the error, whether of excess or defect. Now

Fig. 301.

let us mark off a block, say of 12 inches cube, and we shall find that we must not mark it out all over until one of the faces has been planed up. Suppose, for instance, we mark it out as shown in Fig. 301. The inside lines on faces A and B are the marking-off lines.

If, then, we cut off the metal to the lines on A, we shall have removed the lines on B, and *vice versa;* and there is no manner or means of avoiding the difficulty, save as follows: We may mark off one face, and let the block be cut down to the lines, before marking the other face; or we may have a surfacing cut taken off one face, and then perform the whole of the marking off at one operation. The latter plan is preferable, because it gives us one true face to work from in marking off, and obviates the necessity of having to prevent the rocking of the work upon the marking-off table by the insertion of wedges, which is otherwise very commonly requisite. It is preferable, then, upon all work easily handled and chucked, and in which the lining off must be performed on more than one face, to surface one face before performing the marking out; and supposing our block to have one face so surfaced, we will proceed.

We first well chalk the surface of the work all round about where we expect the lines to come, which is done to make the lines show plainly; we then place the work upon the table with the surfaced face downward; and placing a rule alongside of it, we set the scriber of the surface gauge so as to take off the necessary amount from the top, as shown

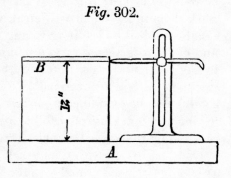

Fig. 302.

in Fig. 302 (A being the plate), and mark the line, B, around all four faces of the work. We then turn the work on the plate so that the true face stands perpendicularly, setting it true by wedging it, so that a square being placed with the back

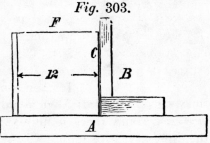

Fig. 303.

to the face of the table, and the blade against the surfaced face of the work, the latter will stand true with the square blade, as shown in Fig. 303. A being the marking-off table, B the square, and C the surfaced face of the work. We then (with the scribing block) mark, across the surfaced face of the work, two lines, 12 inches apart, and of equal distance from the top and bottom faces of the work, as shown in Fig. 304, at A and B. Our next operation is to mark off, on the surfaced face of the work, two more lines, standing at right angles to the lines A and B in the above figure; so that the surfaced face will have four lines upon it. These last two lines we mark without moving the work, by placing a square with its back resting upon the table, the square blade standing vertically and at the necessary distance from the edge of the block, as shown in Fig. 305, A and B being the lines drawn by the scribing block, and C C the square in position to draw one of the necessary perpendicular lines, the other, shown at D, being supposed to have been marked from the square while it was turned around. Here, then, we have the lines for four of the faces, marked upon a face already surfaced to the size, thus disposing of five out of the six faces: and since the line for the sixth face stands diametrically opposite to the surfaced face, the latter has only to be kept down evenly upon the table of the planer to insure the sixth face being cut true; after which, and when each of the remaining

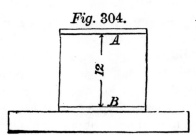

Fig. 304.

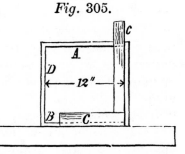

Fig. 305.

four sides is chucked to be operated on, we have a true face to place next to the angle plate, and a true one against which to apply the square to test if the work is held true. Thus we find that the surfaced face of the work, when placed on the face of the marking-off table and on the face of the planer table, becomes a gauge by which (with the aid of the square) all the other faces may be marked or cut true.

It is obvious that, had either one of the faces of the work been faulty, we might have taken off it as much metal as possible, leaving only sufficient to clean up the face diametrically opposite. It often happens that an apparently faulty face shows to more disadvantage by having a cut taken off it; especially is this the case in iron castings, in which there may be more air holes beneath than upon the surface, which defect may be sufficiently serious to spoil the work. It is therefore preferable to take the first or surfacing cut off the defective face, so that the degree of defect may be discovered before even the marking-out is performed.

The lines being marked, our next procedure is to make light centrepunch marks at short intervals along them, so that, if the lines become obliterated through handling the work, the centrepunch dots will serve instead. These dots should be marked very true with the lines, otherwise they destroy the truth of the marking; because the machine operator, in setting the work in the machine, is usually guided by the dots.

By this method we may mark off any body whose outline is composed of straight lines, and whose diametrically opposite faces are parallel, no matter what the length, breadth, and thickness of the body may be. It is not, however, at all times desirable to perform all the marking-out at one operation. Suppose, for example, our work had been a piece of metal 1 foot square and $\frac{3}{8}$ of an inch thick: were we to face off one of the broad faces before marking-

LINING OR MARKING OUT WORK.   373

off, we should find it very difficult to set our work upon the rough edge, and set it true to the square, as shown in Fig. 303; whereas, were we to face off one of the edges first, we have ⅜ of an inch only against which to try the square when setting the planed edge perpendicular. In such a case, therefore, it is best not to mark off the edges until the body of the work is cut to the required thickness.

To mark off a body such as shown at B in Fig. 306, which represents an engine guide bar, one face must either be first trued up, or the marking-off must be performed at two separate operations. The better plan is for the marker-off to examine the bar as to size, and have one face planed off. If either face appears defective, it should be the first planed. If the bar appears sound all over, an outside edge face of the bar should be the one to be planed off preparatory to marking-off; and in setting it to surface it, care should be taken to set it true with the top and bottom faces, if they are parallel to each other; and if not, to divide whatever difference there may be between them. The bar may then be placed upon the marking-off table in the position shown in Fig. 306, A being the marking-off

Fig. 306.

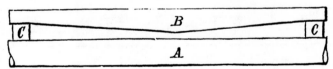

plate, B the guide bar, C C pieces of wood to lift the bar off the plate. By means of small thin wedges, the planed face, B, of the bar, is set at a true right angle to the surface of the plate, and tested by a square. The next operation is to mark off the top or uppermost face, and the question here arises: Shall it be so marked that there will be an equal amount of metal taken off the top and bottom faces, or otherwise? First, then, since the quality of the metal is the best towards the surface, it is a consideration to take off as little as possible, so as to leave a hard wearing surface; this may appear a small matter, but it is always right to gain every superiority attainable without cost. Therefore, all other things being equal, we should prefer to take as little metal off the top face as would be sufficient to make it true, and should therefore mark it out with that view. Here, however, another consideration arises, which is that the outline of the bottom face is not straight, and cannot therefore be planed lengthways from the centre of the bar to the ends at one chucking, and if such bottom face is to be shaped across its breadth, instead of lengthways, it is a comparatively slow operation, and much time will be saved by so marking off the bar that the bottom will only just true up, so that all the surplus metal will be cut off the top face, which, being done in a larger machine, and lengthways, is a much more rapid operation. There is, however, a method of obtaining both the advantage of taking as little as possible off the top face, and planing the bottom face for the most part lengthways. It is shown in Fig. 307, A being the bar; the two

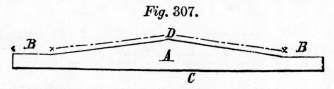

Fig. 307.

faces, B B, may be first planed parallel (as required) with the face, C; the back of the bar may then be planed in two operations from the point, D, to the junction with B at each end. Were this method of procedure employed, it would pay to leave the most metal to come off the back of the bar; but there are yet other considerations, which are the facilities in the shop. If the shaping machines are not kept fully occupied, while the planing machines

are always in demand, it will pay (if there are not many
bars to be planed) to leave as little as needs be to be taken
off the bottom of the bar and the remainder off the top.
If, however, many bars are to be planed, the most economical of all methods will be to plane the backs by placing,
say 8 of them at a time across the table of the planer,
cutting off the ends at the same chucking. Supposing
this plan to be adopted, we set the scriber of the marking
block just below the lowest part of the surface of the bar,
and draw a line along its planed surface, and then another
line along each end, to denote the thickness of the parallel
parts at each end, making this line longer than is necessary, as a guide in setting the bar in the shaper (in case
the ends are shaped and not planed). We next mark off
the length of the bar at the ends, using a square and allowing about an equal amount to be taken off each end;
and then, still using the square, we mark a line equidistant between the end lines to denote the centre of the
length of the bar, which will then present the appearance
shown in Fig. 308, the inside line, A A, being for the top

Fig. 308.

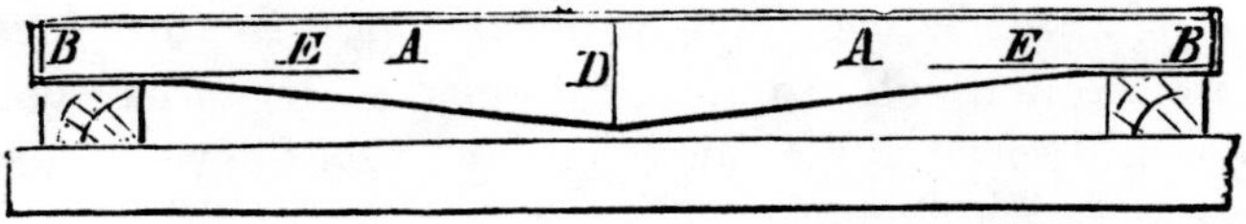

face, the lines, E, for the parallel ends, the lines, B B, for
the ends, and the line, D, denoting the middle of the
length of the bar. We now turn the bar so that its
planed face is uppermost; and, setting a pair of compasses
to the required thickness of the middle of the bar, we set
one point at the junction of the lines, A and D, mark off
with the other point a half circle, and then (turning the
bar over) adjust it upon the table, as shown in Fig. 310,
A being the table, and B a block of wood and wedge to
adjust the bar, so that, if the scribing block be applied

along the table, the needle or scriber point will mark just fair with the top of the circle at D and the mark, C, at the end of the taper part of the bar (the mark, C, showing

Fig. 310.

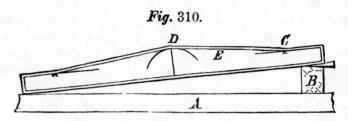

the required distance from the end of the bar). Having made the adjustment, we draw the line, E, thus completing the marking of that half of the bar. We next remove the block of wood and wedge to the other end of the bar, and repeat the last operation, when the marking of the bar will be, so far as its outline is concerned, complete. It will be observed that we have drawn the lines in each case on the one planed surface of the bar only, and not all around the work. The reason for this is that the planed face is a guide, whereby to chuck the work and ensure its being set true. In the absence of one true face it would be necessary, in marking off the first face, to mark the lines all around the work, which, when planed up, would serve as a guide whereby to set the work during the successive chuckings.

Fig. 311.

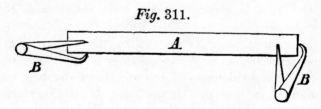

After the faces and ends are planed up, the holes in the ends may be marked by the compass calipers and compasses, as shown in Fig. 311, A being the bar, and B B

LINING OR MARKING OUT WORK. 377

the compass calipers set to the required distance. At the junction of the marks thus made, we make a light centrepunch mark, and mark off the circles for the holes, first marking a circle of the requisite size and defining its outline by other light centrepunch marks. We next draw from the same centre a circle smaller in diameter, and define its outline also by small centrepunch marks; after which we take a large centrepunch, and make a deep indentation in the centre of the circle, which will appear as shown in Fig. 312. The philosophy of marking the holes in this manner is as follows: If the outside circle alone is marked, there is nothing to guide the eye during the operation of drilling the holes (in determining whether the drill is cutting the holes true to the marks or not) until the drill has cut a recess nearly approaching the size of the circle marked; if the drill is not cutting true to the marks, and the drawing chisel is employed, it will often happen that, after the first operation of drawing, the drill may not yet cut quite true to the marks; and it having entered the metal to its full diameter, there is no longer any guide to determine if the hole is being made true to the circle or not. By introducing the inside circle, however, we are enabled to use the drawing chisel, and therefore to adjust the position of the hole during the earlier part of the operation; so that the hole being cut is made nearly if not quite true before the cutting approaches the outer circle, which shows the full size of the hole. If, on nearly attaining its full diameter, the outer circle shows it to be a little out of truth, the correction is easily made. It is furthermore much more easy to draw the drill when it has only entered the metal to, say, half its diameter than when it has entered to nearly its full diameter.

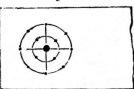

Fig. 312.

The object of making a large centrepunch mark in the centre is to guide the centre of the drill, and to enable the operator to readily perceive if the work is so set that the point of the drill stands directly over the centrepunch mark. This is of great importance in holes of any size whatever, but more especially in those of small diameter, say, for instance, ¼ inch, because it is impracticable to describe circles of so small a diameter whereby to adjust the drilling; and in these cases, if the drill runs out at all, there is but little practical remedy. The centrepunch marks for such holes should therefore be made quite deep, so that the point of the drill will be well guided and steadied from the moment it comes into contact with the metal, in which case it is not likely to run to one side at all. If a motion or guide bar requires to have one corner rounded off, as it should have to prevent its leaving a square corner on the guide block, which would weaken the flange of the latter, the corner cannot be marked off, but a gauge should be made as shown in Figs. 313 and 314,

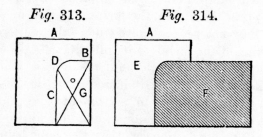

Fig. 313.   Fig. 314.

A in Fig. 313 being a piece of sheet-iron, say $\frac{1}{32}$ inch thick, with the lines, B and C, and the quarter circle, D, marked upon its surface. The metal, G, is then cut away, and the edges carefully filed to the lines, thus forming the gauge, A, which is shown upon the bar, F, in the position in which it is applied when in use. It is obvious that such a gauge will scarcely suffice to get up a very true round corner; this, however, is accomplished by leaving the corner of the work a little full to the gauge and then filing it up to the piece of work fitting against it.

TO MARK OFF THE DISTANCE BETWEEN THE CENTRES OF TWO HUBS OF UNEQUAL HEIGHT.

When the heights of two hubs are unequal, as shown in Fig. 315, the distance required being that from A to B, we must make the necessary allowance (in the distance at

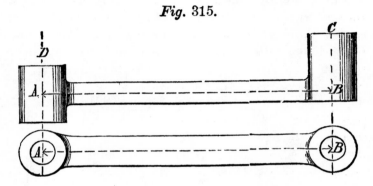

Fig. 315.

which we set the compass or trammel points) for the difference in height of the surfaces upon which our circles are to be marked, from the body of the lever or arm. If

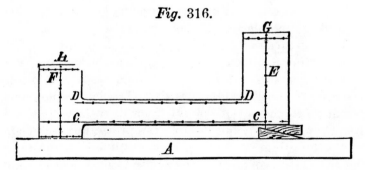

Fig. 316.

the arm is to be finished along its whole length, it is better to mark off the body of the arm first, which we perform as shown in Fig. 316. Setting our work upon the table A.

and wedging it as shown, we mark off with the scribing block the lines C C and D D, making their distance apart the thickness of stem required, and leaving about an equal amount of metal to be taken off each face. We then mark off the height of each hub face, measuring from the line C, and scribe a line around each hub face as far as the scriber point will allow. We next mark off (with a square, resting against the surface of the marking-off table) the lines E and F, marking them as near the centre of the hub as the eye will direct, their use being simply as guides in setting the work in the lathe or machine. These lines being dotted with a fine centrepunch to prevent their be-

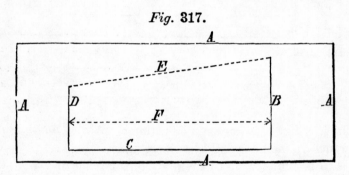

Fig. 317.

coming obliterated, we next measure the height of the face G, and that of the face H, both from the line C.

We now turn to the marking-off table, and on its surface draw a straight line a little longer than the length of our arm or lever, as shown in Fig. 317, the lines A A A A representing the outline of the marking-off table, the line B representing the height of the hub from its surface, G, to the line C, in Fig. 317, and the line B representing the height of the hub from its surface, H, to the line C in the same figure. The two lines B and D are to be struck at right angles to the line C, and the distance between them (as denoted by the dotted line F) being the required distance from centre to centre of our lever. These lines being drawn, we have only to set our compass or trammel points to the length of the dotted line E, to be able to mark off the correct distance apart for the centres of the circles to be marked on the faces of the two hubs. Proceeding, then, we place our lever on the marking-off table in the position shown in Fig. 318; and after putting a centre-

Fig. 318.

piece in each hole, we draw (along the entire length of the lever, and across the faces of the hubs) the centre line A, locating it in the centre of the stem. We then apply the trammels, set as already directed, to mark off the centres of the holes. Setting our compasses at the intersection of the line A with the line marked on each of the hub faces, we strike the necessary circles on the faces of the hubs, as shown. We next mark off the breadth of the lever or arm on the face from the centre line A, and our marking is complete.

MARKING HOLES AT A RIGHT ANGLE.

To mark off a crosshead in which one hole requires to be at right angles to the other, we proceed as follows: First placing the crosshead upon the marking table, as in Fig. 319, we draw with the scribing block the centre line A, marking it all round the crosshead; and if the crosshead has a hole or holes in it, we put centre-pieces in those holes to receive the centre lines. We then place a square with its back resting upon the marking-off table, and draw, parallel with the edge of the blade, the centre line B. From the intersection of the lines A and B we draw the lines C and D, marking their distances from the line A with a pair of compasses, and carrying the lines round

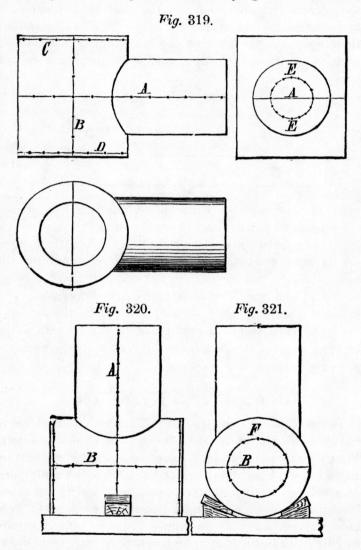

Fig. 319.

Fig. 320.   Fig. 321.

with the scribing block. We draw the circle E (Fig. 319), using the line A as a centre, and locating it, as nearly true as we can (the other way), to the hub or stem. We now stand our crosshead in the position shown in Figs. 320 and 321, and applying a square to the line A, we set it to a right angle with the face of the line A, wedging it upright with the wedges shown. Then, setting the scribing block needle point even with the line B of Fig. 319, and setting that line true with the surface of the table, we carry it across the other face, as shown in Fig. 321, locating its position sideways to suit the forging or casting, and then we strike the circle F, which completes the marking.

It will be noted that the lines A and B are mere guides whereby to obtain the centres of the circles from, and it may therefore be asked for what purpose those lines are centrepunch-marked. The reply is, that those lines must be used as guides to set the work by when chucking the crosshead on the lathe or machine.

TO LINE OUT A DOUBLE EYE.

After measuring the dimensions of a double eye to ascertain if there is, upon the outline, surplus metal sufficient to permit of its clearing up all over, we apply an L square upon the outside surfaces, and a T square, with the blade between the jaws, to test if the inside and outside faces are at about a right angle to each other, or if the marking will have to be thrown to one side of the work to accommodate a want of truth in the latter. Presuming that, as is usually the case, the work is reasonably near to being true, we proceed as follows: Placing the double eye upon the marking-off table, as shown in Fig. 322, we block up the stem end with the pieces of wood, B, so that the horizontal faces of the work will stand about true with the surface of the table, the manner of testing the same being shown as applied to a square block in Fig. 323, A representing the marking-off table, and B the scribing block with the needle placed on a piece of work so that the point of the bent end barely touches the surface of the work. The operation is to move the scribing block from end to end of the work and on both sides of it, packing it up until the upper surface is level, and taking care, if the work

Fig. 322.

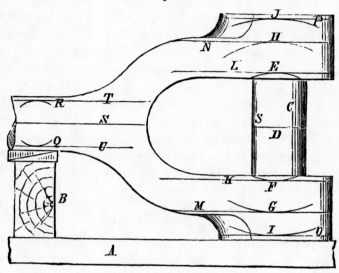

Fig. 323.

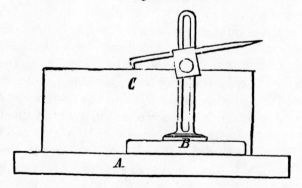

does not lie level and steady upon the table, to insert wedges in the necessary places so that the work will lie firmly and not move during the operation of marking.

LINING OR MARKING OUT WORK.

If there are projections upon the face of the work which rests upon the table, as is the case in our double eye, it is necessary to pass the scribing block along the under as well as the upper surface of the work; and if the two vary much, we may choose the one that is most true with the other surfaces of the work and set it true; or if, in such case, there would not be enough metal to clean up the work on both sides, we must divide the difference between the two. We then put between the jaws of the double eye, the centre piece, C, Fig. 322, and find the centre of the jaws, as shown by D; then, setting a pair of compasses to half the required width between the jaws, we scribe upon both the jaws the segments of a circle, E and F, using D as a centre; then opening the compasses to allow for the requisite thickness of each jaw, we mark the segments of a circle, G and H; and again setting the compasses to the requisite thickness of hub, we mark the segments of a circle, I and J. We now take a scribing block, and, setting the point just to intersect the extreme diameter in each case, we draw the lines, K and L, M and N, and O and P, thus defining the widths and thicknesses of the jaws and hubs. We then set the scriber point even with the centre, D, and then draw the line, S S, which should run a long way up the stem of the double eye, because the shortness of the other lines, running parallel to it, renders it difficult to set the work true by them, and S S is made long to supply the deficiency. After setting the compasses to half the required thickness of the stem, we mark, using the line, S S, as a centre, the segments of a circle, Q and R, and from them mark the lines, T and U, which define the required thickness of the stem or rod of the double eye. Our next operation will be to mark off the hole and the circle of the hub, which is done as shown in Figs. 324 and 325. Setting the eye upon the marking-off table, A, we wedge it upright, as shown in the side view, Fig. 324. by the wedges, B; applying the blade of an L square to set the line, S S, Fig. 325, true by, we mark off on each side of the double eye the centre

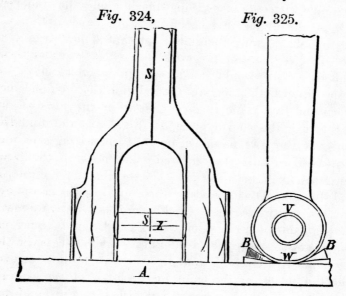

Fig. 324.   Fig. 325.

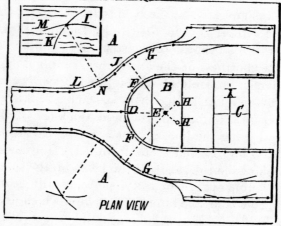

Fig. 326.

PLAN VIEW

of the boss or eye, and from that mark off the circles, V and W, denoting the finished sizes of the hole and the

eye; then setting the scribing block needle point even with the centre from which the circles, V and W, were struck, we mark on the centre piece (shown in Fig. 325) the line, X.

We have now to complete the marking-off of the edge face shown in view 2, Fig. 325, which could not have been done before, because there was nothing determinate wherefrom to mark off the half circle of the outline between the jaws. Placing the double eye upon the table, as shown in Fig. 326, and blocking it up so that it lies level with the face of the marking-off table, and with the face that has been marked off, uppermost, we insert between the jaws the centre piece, B. We next mark from the centre, C, the requisite distance from the hole in the eye to the crown of the curve, between the jaws, thus obtaining the centre mark, D, from the centre, C; and setting the compasses to half the required width between the jaws, we use D as a centre, and mark upon the centre piece, B, the centre, E, and then strike the half circle, F F, which completes the marking between the jaws. Our next procedure is to mark off the segments of circles, G, G, which are struck from the centres, H, H, respectively. Then taking the block of wood, I, which should stand at about the same height from the marking-off table as does the body of the double eye, and setting the compass to the required radius, we rest one point on the circle, G, at about the point, J, we strike the mark, K; then placing one leg of the compasses at about the point, L, we strike the line, M, the junction of the lines, K and M, forming the location of the centre from which the segment of a circle, N, is marked. Placing the block of wood, I, on the other side of the double eye, we repeat this latter operation, and the marking on that face is complete.

After defining the outline of our work by light centre punch marks, we pass it to the machinist's hands to be turned and cut down to the lines, after which we place it upon the marking-off table in the position shown in Fig. 327, A representing the table. At each side of the double eye we place in the hole a centre piece, B, and mark

Fig. 327.

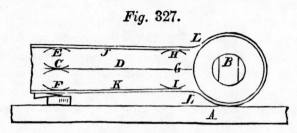

thereon the centre of the hole with the compass calipers. We then find the centre of the shank, C, and, wedging that end up with wood as shown, we set the needle of the scribing block even with the centre of the hole, and so adjust the double eye with wedges that the needle point will strike the centre of the hole marked on B, on each side, and also the centre, C, whereupon we may mark the line, D; then setting the compasses to the requisite distance, we mark from the centre, C, the segments of circles, E and F, and resetting the compasses on account of the taper from the centre, G, the segments of circles, H and I: and resetting the double eye so that the needle point of the scribing block will intersect the extreme outline of H and E, we draw the line, J; repeating the operation on the under side, we produce the line, K, and the operation is complete. The curves, L L, are made to a gauge, such as is shown in Fig. 328; it is made of sheet-iron about one-sixteenth of an inch thick, the outline being carefully marked out and filed up neatly, the corner, A, being made of the necessary sweep, and the hole, B, being used to hang the gauge up by. It

Fig. 328.

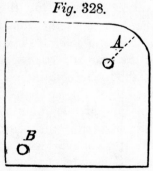

is well to have an assortment of such gauges for use in lining out, as well as for use as guides to the machinist in cutting out the curves or sweeps.

To assist the operator in marking out, the centres from which all curves and circles on detail drawings are struck should have a small circle in red ink marked around them, and a dotted red line marked from the centre to the circle or segment of circle struck from it, similar to the dotted straight lines shown in Fig. 326. If the double eye is, however, intended to have an offset, as

Fig. 329.

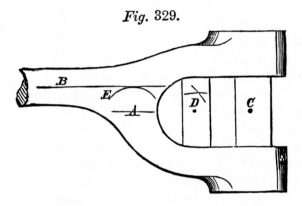

shown in Fig. 329, we draw from the centres, C and D, the line, A; and setting the compasses to the amount of the offset, we draw the segment of a circle, E, using the line, A, as a centre; and from the extremity of that segment, we draw with the scribing block the line, B, which will represent the centre line of the stem of the double eye, the rest of the operation being as shown in Fig. 322, and described in the accompanying explanation, from the point at which the line, S S, in that figure was drawn.

MARKING OUT AN ECCENTRIC.

In measuring an eccentric to ascertain if it has sufficient stock to allow it to be cleaned up all over, it is not sufficient to measure the thickness, the outside diameter, and the size of the bore only, because those measurements do not take the amount of the throw into consideration, and we have, therefore, to proceed as follows:

In Fig. 330 let A A represent an eccentric, into the bore of which, on the hub side, we place the centre piece, and mark upon it the centre of the hole. We then take a pair of compasses, and set them so that, when one point is resting in the centre of the hole, the other point will reach to within about a quarter of an inch of the extreme diameter of the eccentric, as shown above by the line, C C.

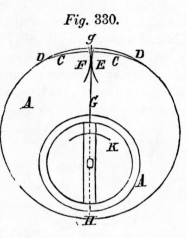

Fig. 330.

We then, with a pair of compass calipers, find the centre of the line, C C, by resting the caliper leg of the same against the periphery of the eccentric, at one of the points where the line, C C, meets it; and then with the compass leg of the compass calipers, we mark the line, E; and repeating the operation at the other end of the line, C C, we mark the line, F. We next take a straight edge; and placing it so that its edge is even with the centre of the bore of the eccentric and with the centre between the lines, E and F, we draw the line, G g, upon which we may make our measurements as follows: After setting a pair of compasses to the amount of throw required by the eccentric, we place one compass leg in the centre of the bore, and with the other mark (on the line, G) the line, K, which will represent, at its intersection with G, the centre of the finished diameter of the eccentric. Then we take a rule, and measure from the centre, K, to the ends, H g, of the line, G, which ends should be equidistant from K, if the amount to come off

LINING OR MARKING OUT WORK. 391

the surface of the casting in the hole is to equal that to come off the outside surface. It very frequently happens, however, that there will be more to come off the eccentric on one side of the diameter than on the other, especially when the eccentric is put together in two halves; because, in facing up the two halves, preparatory to putting them together, and to make them bed well one to the other, it does not always happen that the same amount of metal is taken off each face. Again, the quantity so taken off is not always that allowed on the pattern for the purpose; so that, in practice, an eccentric casting rarely marks off true with its rough outline.

Here, then, arises the consideration as to in what direction we shall throw the lines. Shall it be to bore the hole true, or to turn the outside diameter true, with the casting? The latter plan is always preferable; because, if in turning up the outside diameter the first cut does not true it up, the tool point will scrape over the sand, after leaving the cut and before it strikes it again, to such an extent as to rapidly destroy the cutting edge, necessitating not only frequent regrinding of the tool, but also that its cutting speed be very materially reduced. After having roughly ascertained, in the manner described (which process will take but a few minutes to perform), that there is surplus metal enough to clean up the eccentric, we may proceed to mark it out.

It is much easier to mark off an eccentric on its plain side than on the side on which the hub stands, because of the projection of the hub; and, furthermore, the marking for the hole and for the diameter can be performed at one operation, which is impracticable on the hub side. But if this plan is not adopted, it necessitates that, at the first chucking, either the hole only shall be bored, in which case there will be no face true with the hole, and hence no guide whereby to set the eccentric at the next chucking: or else, in turning off the outside face after the hole is bored, the marks for the second chucking will be effaced. The main consideration, however, is that there is only one way to chuck an eccentric to insure its being turned as true as possible; and the marking off must, therefore, be made to accommodate the chucking, the method and reasons for which are explained in the remarks on eccentrics, on page 136.

The lines to mark the location of the hole and the thickness of the hub may be marked in the manner shown in Fig. 330 or we may adopt the plan shown in Fig. 331, which is perhaps the better of the two.

From the four points A, B, C, and D, we mark off, on the hub side of the eccentric, the centre of its diameter, E; we then, setting a pair of compasses to the amount of throw required for the eccentric, mark off from the centre, E, the arc, F; then, with a pair of compass calipers, placed in each case with one end against the bore of the casting,

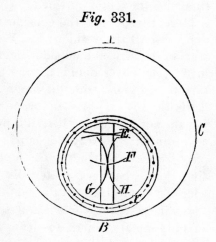

Fig. 331.

we mark the arcs, G and H, the junction of the arcs, F, G, and H, being the required centre of the hole. We therefore strike from that centre, around the face of the hub, the line, I, and mark it lightly with a centrepunch, as shown. If, however, it should be found that there is not sufficient metal to allow the hole to be cleaned up if marked off true with the circumference, we must throw the hole a little in the requisite direction, endeavoring (for the reasons already stated) to keep the diameter of the eccentric as nearly true for the throw as possible. For

LINING OR MARKING OUT WORK. 393

instance, in Fig. 332, if we suppose there is an insufficiency of metal in the hole, at A (E being the centre of the diameter of the eccentric, and K the amount of the throw), we first set a pair of compasses to the required radius of the diameter; and from the centre, E, strike the circle, F, which will show the amount of metal required to be taken off the circumference of the work, and therefore to what degree we are able to throw the hole to accommodate the scant spot, A. If there is more metal between the line, F, and the periphery than the spot, A, lacks, the eccentric will clean up, and we may mark off the hole, allowing it to just clean up, as shown by the circle, L. It is, however, best, on small eccentrics, to mark the circle, L, as large as the face on which it is marked will admit; because, the larger the circle, the less a slight want of truth in the chucking will affect the truth of the work. It will be observed that, in consequence of the centrepiece standing above the level of the face (to the amount of the height of the hub), the circle, F, in Fig. 331, would be too small if marked with the compasses set to the correct radius; but since the duty of that circle is to merely indicate the amount of surplus metal on the outside diameter, it will be sufficiently correct on ordinary eccentrics, to mark it as directed, making a slight allowance of increase in setting the compasses to draw that circle. If, however, it should happen that the quantity of stock is so scant as to make it questionable whether the work will true up: then the centre piece may be lowered in the hole to the level of the surface of the metal

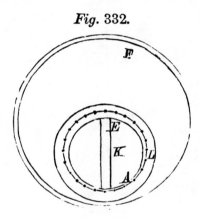

Fig. 332.

on which the circle, L, is marked, and the compasses may be set to the correct radius.

The hole being marked, no further marking should be performed until the eccentric has had both sides finished and the hole bored, when the diameter should be marked upon the plain side of the work, as shown in Fig. 333. After inserting the centre piece, and marking off upon it the exact centre of the hole, we mark the line, C C; and finding the centre of its length, as already described, we strike the line, D; then we mark on the line, D, the amount of the throw, measuring from the centre of the hole, and we thus obtain the centre, F, from which we mark the circle, G G, which is only intended to be employed in setting the work, and need not, therefore, be made of any particular size. The marking will thus be completed, and it will be noted that the thickness of the eccentric and the hub, and the height of the latter, have not been dealt with at all, the reason for the omission being that it is entirely unnecessary to regard them, since (providing of course that there is spare metal enough to clean them up) they may safely be left to the turner, who may accommodate the amount taken off the first side faced, according to the smoothness of the second cut, or a variety of other conditions which need not be here enumerated. If the eccentric has no hub, as is sometimes the case, it should be marked off, as shown in Fig. 331.

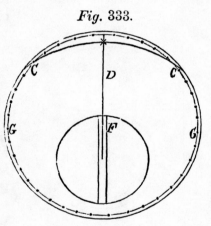

Fig. 333.

After the turning is completed, the keyway or featherway may be marked off, as shown in Fig. 334. Placing the centre piece on the hub side of the eccentric, so that the plain side may lie flat on the slotting machine table, and not require parallel strips or packing wherewith to chuck it, we mark off upon it the centre of the hole in the eccentric; and from that centre, we mark a circle whose diameter must be equal to the required width of the keyway to be cut. Then selecting the location of the keyway, we describe there another circle of the same diameter as at A. Placing a straight edge so that one of its edges is just even with one and the same side of each circle, we draw the line, A; and by repeating the operation on the other side of the circle, we shall have the sides of the keyway marked. To mark the depth, we make a fine centrepunch mark at the requisite distance from the bore of the eccentric, and then, using a square place one of its edges parallel with the outer edges of the two circles, and the other edge fair with the centre of the centrepunch mark, and scribe a line along the latter edge and across the width of the keyway, the operation being shown in Fig. 335, A being the square. When, however,

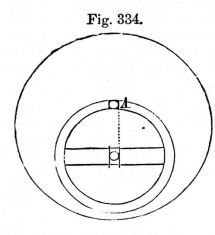

Fig. 334.

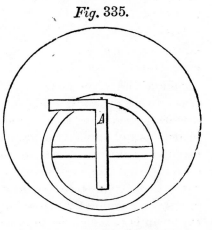

Fig. 335.

there are a number of keyways of the same width and depth to be marked, it is more expeditious to make the gauge shown (together with its method of application) in Fig. 336, in which A represents the gauge, being a piece of sheet-iron about one-sixteenth of an inch thick, the curved line being of the same curvature as the bore of the hole in the eccentric, and the projection, B, being of the required size of keyway. The ends, C D, are to be slightly bent (both in one direction), so that, while the projection, B, will lie on the face of the hub, the ends, C D (being depressed), will contact with the bore of the hole of the

Fig. 336.

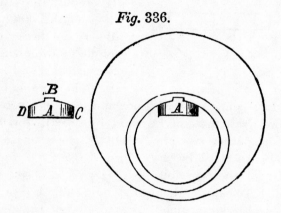

eccentric and thus serve to keep the gauge true with the bore. The gauge should be carefully marked out and smoothly filed true to the lines. The small hole, shown near C, is to hang up the gauge by when it is not in use.

LINING OUT CONNECTING RODS.

Connecting rods, so large in size as to be cumbrous to handle, are generally made by forging the ends to which the strap is attached by themselves, and afterwards welding them to the body of the rod: the advantage being that the machine work done to the rod ends can, in that case, be done in small machines and at a higher rate of

LINING OR MARKING OUT WORK.

cutting speed than would be possible if, the rod being solid, its whole body had to be chucked in order to operate on the ends only. If any finishing is required to the body of the rod, it is in such case done after the rod ends are welded to it and made true to the already finished block end of the rod. If, however, the rod is forged solid, the whole of the marking-off should be gauged to suit the body of the rod. For instance: If the stem of the rod is round, the marking-off of the ends should be performed from a centre marked off true with the round stem and on the end face of the rod. The first operation should in this case be, after marking off the said centre, to put the rod in the lathe and face off the block end faces, thus giving us a face, at each end of the rod, true with the stem of the rod, and therefore useful not only to receive the marking off lines but also as a face whereby to true the other faces on the block or stub end. If the ends are forged separately from the body of the rod, it is better to face off one of the side faces, and to mark off on that side face. To mark off a rod end that is forged solid with the stem of the rod, we proceed as shown in Fig. 337, A representing the centre, true with the body of the rod; B B shows the diameter of the rod end struck with the compasses from the centre, A, and C C, the thickness of the rod struck in like manner. If there should not be sufficient metal on the block end to permit the marking-off to be performed from the centre, A, when true with the body of the rod, that centre must be moved sufficiently to allow the rod end to be cleaned up; this is, however, to be avoided if possible, for the following reasons: If the body of the rod runs much out of true, the turning of it in the lathe will be a slow process, because such rods are liable, from their length, to spring in consequence of the pressure of the cut. Hence it is not practicable to take heavy cuts along

Fig. 337.

it; and if in consequence of the body of the rod running much out of true, it cannot be cleaned up at one cut, the tool will scrape, during the first cut, against the scale, necessitating that the cutting speed of the tool be much less than it otherwise need be.

After the segment of circles, B B and C C, in Fig. 337, are struck, which may be done before setting the rod on the marking-off table, the rod should be set on the marking-off table with one of the broad faces downwards, and with the scribing block needle point placed level with the mark, C, on the upper face; and the rod should be tried along that face to ascertain if there is sufficient metal to clean it up all across. The scribing block should then be carried to the other end of the rod, and tried with the upper mark, C; and that being found correct, the scriber point should be set to the lower mark, C, at each end of the rod; and thus the two lines across the rod end, representing the thickness thereof, may be drawn by the scribing block at each end of the rod. The lines representing the breadth of the block end of the rod may then be drawn by simply placing a square on the surface table, with the edge of the square placed in each case level with the extreme diameter of the segments of circles, B B, Fig. 337. No other lines in this case will be required, because the rod ends, having been turned in the lathe, give the machinist two true faces whereby to set the rod at each chucking. If the rod ends are not welded to the rod, the better plan is to have one of the broad surfaces on each rod end surfaced up in a planing machine, and to then perform the marking out on the surfaced faces. The marking out should be made about true with the stem of the rod, as shown in Fig. 338. The surfaced face is to be set, by a square, to a right angle to the marking-off table face; and the centre line, A A, of the stem is found from the body of the stem, and carried from end to end of the **forging** as a guide to set the work by, the lines, B B or

C C, being too short to serve the purpose. These latter lines are struck equidistant from A A. The line, D, should be struck with a square resting on the marking table, and any surplus metal should be taken off the end face rather than out of the corner where the butt joins the stem; because it is easier to take the metal off the end than out of the shoulder. The round corners need not be marked, it being preferable to make a gauge to shape them to. The edges thus marked being shaped off, the thickness of the butt end may be marked off by a scribing block, the planed surface of the butt end lying flat on the marking table. The strap should first have one face

Fig. 338.

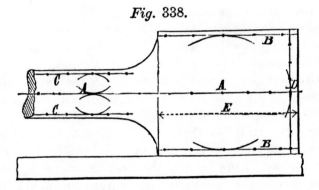

surfaced, and then a centre-piece should be placed between the jaws, being made just sufficiently tight to be held, and *not so tightly as to sensibly spring the jaws open;* otherwise, while the thickness of the jaws would be marked off correctly, the width between them and their outside diameter would be too small when finished. The strap should then be placed on the marking table, and marked as shown in Fig. 339, the lines A A, B B, and C C, being marked off to the required widths apart and equidistant from the centre line, marked across the centre-piece and across the crown of the strap, at D E. The centre of the centre-piece having been obtained from the inside of the jaws, and carried across, at D E, after the strap is set upon the table with the inside faces of the jaws parallel with the face of the table, the width between the lines, A A, should be marked less than is the width of the block end on which they fit, for the following reasons: A connecting rod strap will, by reason of its shape, spring open between its jaws very easily indeed; and were the width between the jaws made the same as that of the block end of the rod, the strap would fit very loosely to its place. It is therefore necessary to make allowance for this in the width between the jaws of the strap, making them narrower than the block end of the rod. The amount of this allowance depends upon the size and stoutness of the strap, an ordinary proportion being about one-sixteenth of an inch to a strap eight inches wide between the jaws. This amount of allowance will enable the strap to spring over the rod end, and be a good fit; that is to say, not so tight but that it can be easily pulled off by the hand, and not so loose as to fall off of its own weight if unsupported. Then, again, any ordinary amount of metal removed in fitting the strap to the rod end will not seriously affect their fit together. Now it is obvious that, if the rod end faces on which the jaws of the strap fit are made parallel to each other, the strap, in being sprung on, would spring open so that its jaws would only touch the block at its entrance end, the end of the jaws standing open from the block end. To obviate this, the block end faces B B, in Fig. 338, are made slightly taper, that is to say, about one-thirty-second of an inch or rather less in a length of six inches, the diameter of the end being the smaller. It is not necessary to mark so small an amount of taper in the marking, it being sufficient to run the centrepunch dots a little inside the line at the end of the block on each side. The lines, A A, in Fig. 339, representing the inside jaw faces, should also be a little taper, first to allow of fitting the strap to the block end, and next to make the fitting of the brasses into the strap an easier operation. It is obvious that, if the inside jaw faces of the strap are parallel with each other, so soon as the brass is reduced to the size of the top of the strap, it will slide clear down to its bed; whereas, if those faces are made a little wider apart at the open end than at the crown end, the brasses, after entering at the open end, will have metal sufficient to be taken off them (before being let down to the crown) to permit of their being fitted nicely to the strap. For these reasons, the faces of the strap, A A, in Fig. 339, are made wider apart, in the proportion of nearly one-

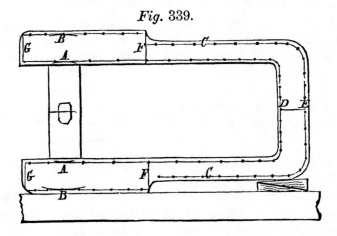

Fig. 339.

sixteenth of an inch of taper to a strap having a jaw twelve inches long. The line, D, in Fig. 339 representing the amount of metal to be cut out of the crown of the strap, should only need that sufficient metal come off to allow that face to just true up: because it is an awkward face to operate on, and it is much easier to take any surplus metal off the outside crown of the strap, as represented by the line, E, in Fig. 339. The lines, F F and G G, are marked at the requisite distance from the crown, D, of the strap, with a square resting on the face of the marking table. The round corners and

34*

curves are marked off with the compasses, using the blocks of wood shown in our lesson on marking off a double eye, previously given. The finishing, however, of such corners, both in the machine and in the vise, is usually done to a small sheet-iron gauge. Such corners may, it is true, be cut on a slotting machine table to a correct sweep without the use of a gauge; and there are many shaping machines with special attachments for the same purpose.

Our next operation is to mark out the keyway, which is performed after the butt end of the rod and the inside and outside of the strap have been planed. We first, with a pair of compass calipers, which are better for the purpose than compasses, mark the centre of the strap edgeways, and then, laying it with its broad surface on

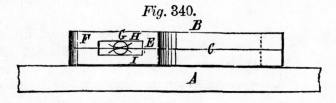

Fig. 340.

the marking-off plate, we mark off the keyway as follows: In Fig. 340 A represents the table, and B the connecting rod strap. C is the centre line of the strap, and therefore of the keyway; the end, E, of the keyway should be drawn the necessary distance from the inside crown of the strap, as denoted by the dotted vertical line, because it is that distance upon which the thickness of the brasses depends. Hence the line, E, is the first one to be drawn; then, from the line, E, we mark the length of the keyway, and strike the line, F; the breadth of the keyway we mark by setting the compasses to the radius of a circle whose diameter will be equal to the required breadth of keyway. Then using the centre line as a centre, we mark the circle, G, and (touching its diameter, and parallel with the centre line) the lines, H and I, thus completing the marking of the keyway on the strap.

To mark off the oil hole, we lay the strap on its side face, as shown in Fig. 341, and, placing a straight edge along the inside crown face of the strap, we mark a line even with it and across the jaw of the strap, as shown at A, and from that we mark with the compasses the line, B, the distance between the two being half the total depth of the brasses, or what is the same thing, the thickness of the crown brass (when new) from its joint face to its bedding crown. We then, with a square and scriber, carry the line, B, over to the centre line of the edges of the strap (C in Fig. 340), and the junction of the two is the centre of the oil hole. In centre-punching the centre for the oil hole to be drilled, make a deep centrepunch mark to prevent the drill from running to one side and thus deceiving the machinist (who may have to line up the brasses when they become worn) as to thickness of the liner to be placed behind the back brass to keep the rod to its original length.

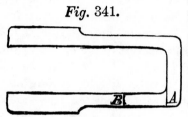

Fig. 341.

The marking of the keyway in the butt or stub end of the rod is performed in the same manner as that of the keyway in the strap, care being taken to make the edge of the keyway nearest to the end of the rod at the exact proper distance from that end: otherwise the amount of space left, when the strap is in its place, between the end of the rod and the crown of the strap (which regulates the thickness of the brasses), will not be correct, and the oil hole will not stand in its correct position on the strap, unless the key and gib are made to suit the inaccuracy of the position of the keyway in the rod end. For example : **Suppose the keyway of the rod to approach too near** the rod end; then the strap will, if the gib and key are made of the proper width across, as at A, not pass sufficiently along the block end, and there will be too much space allowed for the brasses, and the oil hole will stand too near the crown of the strap. The only method of correcting this defect is to make the width of the key and gib, at A, Fig. 342, wider to the necessary amount, and to cut the keyways, both in the strap and the rod end, wider, by cutting out the metal on the edge of the keyway farthest from the rod end, and the metal on the edge of the keyway in the strap at the end nearest to the crown of the strap. If the keyway of the block end errs in the opposite direction, the keyways must of course be made wider, the metal being cut out in the exact opposite to the above direction. By marking out the two keyways as above described, we have no occasion to take any account of the draw, since that will come right of itself when the brasses are put in their places in the strap, and the strap is put in its place upon the rod end. In marking off the keyway in the rod end from keyways already cut in the strap, the following plan must be adopted : Place the strap upon the rod end, leaving the space between the rod end and the crown of the strap narrower than is required to receive the brasses (when the latter are new) by an amount equal to the amount of taper there is in the full length of the key, and mark the keyway in the rod end even with the strap, taking no account of the draw required on the keyway, which is provided for in the position in which the strap is placed on the rod end, as will be perceived when we consider that the length of a keyway is always the width of the key and gib, at A, when placed together, as shown in Fig. 342. Hence, when marking off the keyway in the rod end by the keyway in the strap, the latter should be placed in the position in which it will stand when the key and gib are in the position shown in Fig. 342. Supposing then the gib and key to be in their plac s

LINING OR MARKING OUT WORK.

in the rod and strap, and in the position shown in Fig. 342, and that we then lift the key up so that it will stand in the position shown in Fig. 343, and that we then pull the strap as far off the block end of the rod as it will come, the key will then stand in its correct position, and there will be the proper amount of draw in the keyway, both in the strap and on the rod end, and the space between the end of the rod and the crown of the strap will also be correct. To mark off the key and gib, we proceed as follows: After the keyways are filed out, we take a piece of thin sheet iron and fit it to a tight fit in

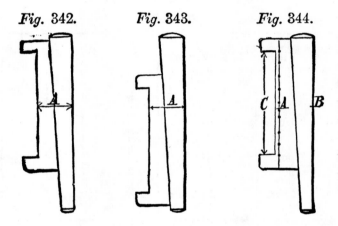

Fig. 342.   Fig. 343.   Fig. 344.

the breadth or thickness of the keyway, and have the thickness of the key and gib planed, using the piece of sheet iron as a gauge; we then mark off the key on both edges to the proper width at top and bottom, and hence give it the correct amount of taper. We also have the plain or straight edge (that is, the edge opposite to the jaws) of the gib planed straight; we then place the gib and key in the position shown in Fig. 344, and mark off (from the edge face, B, of the key) the line, A, on the gib, using the compass calipers set to the full width of the keyway in the strap or rod end, taking no account of the draw.

Hence the key and gib will, when in the position shown, just fill the keyway. The width between the jaws of the gib, as denoted by C, should be marked a trifle less than is the extreme outside width of the jaws of the strap, so as to allow for the metal taken off in filing up the outsides of the jaws of the strap and off the inside of the jaws of the gib.

To mark off cylinder ports and steam valves: Beginning with the cylinder, we place in the exhaust port a centre-piece, as shown in Fig. 345, in which A represents the steam port, B B the cylinder exhaust port, and C the

Fig. 345.

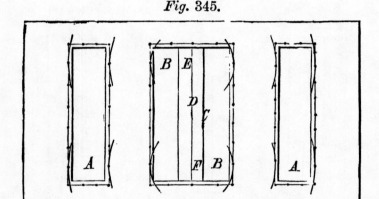

centre-piece wedged or fastened therein. In the centre of the position intended for the ports, we mark upon the centre-piece the centre line, D, and from the points, E, F, we mark with the compasses the segments of circles from which the width of the steam-ports, exhaust port, and bridges are marked, the lines being drawn by the aid of a straight edge. We mark the ends of the ports by the aid of a straight edge and square. To mark off the valve, we may either plane up two of the edges and mark the lines by the aid of a square, allowing an equal amount to be taken off each side of the exhaust port, or we may place a centre-piece in the exhaust port of the valve, and perform all the marking-off before any of the planing is done, the operation being shown in Fig. 346. From A to B is the width of the exhaust port of the valve, and from C to D on each side is the lap of the valve.

It is found that valve seats (the cylinder faces on which the valves slide) will have, when they become worn, a

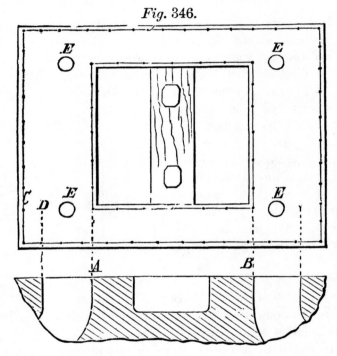

Fig. 346.

groove cut across the bridges between the ports and extending along the face beyond on each side, running close to the edge of the ports, and at right angles to the lengths of the ports. To prevent the formation of this groove, it is found necessary to mark after the face of the valve has been planed the four small holes (say of $\frac{1}{4}$ inch diameter) shown in Fig. 346, at E, E, E, E, their centres coinciding with the edges of the exhaust port of the valve.

To mark off the back of the valve where the slide spindle frame fits, we must stand it on the marking table, with the face standing perpendicularly and at a right angle to the face of the table, and draw a centre line on the back of the valve, from which line we may mark off the back of the valve to the necessary conformation.

TO MARK OUT A CONE PULLEY.

In laying out cone pulleys, or stepped cones, as they are sometimes termed, for crossed belts, the belt will have equal tension when placed to run on any of the corresponding steps of the cones, providing that their axial shafts are parallel one to the other; that the largest cone of one pulley is in line, or *fair*, with the smallest cone of the other; and that the steps of the cones are equal, so that a line drawn from the centre of the faces of the two end steps passes also through the centre of the faces of all the other steps, as shown in Fig. 347, by the lines, A and B. It fol-

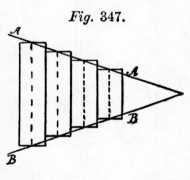

Fig. 347.

lows also that if the cones were plain, that is, without steps, the tension would remain equal with the belt at any location on the cone.

In Fig. 348, A and B represent two stepped cones connected by the two belts, C and D, the latter running upon the middle, and the former on the end cones, the belts being crossed. It will be noted that in consequence of the crossing of the belt it has contact, on both the top and bottom steps, with more than one-half the circumference of the steps, but the variation this would cause in the length of the belt is too slight to be of any practical value; hence it may be discarded. Measuring then the lengths of each side of the belt from the centre-line **I I** to the centre-line **J J**, we shall find D and C to be equal; and if we assume the diameter of the step marked 1 to be 10 inches, and that of the smallest, marked 3, to be 4 inches, the half of their added circumferences will be 21·99. In this case, the sizes of the middle cones, marked

Fig. 348.

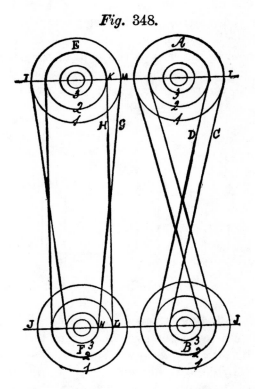

2, will be 7 inches, and the half of their added circumferences (and this is the part of the cones around which the belt would lap) will also be 21·99 inches; hence, the sum of the arcs of contact of the belt around the pulley being equal in both cases, and the lengths of the sides of the belt being equal, it is obvious that the tension of the belt will also remain equal.

If the two cones are to be connected by an open or uncrossed belt the case is different, because the length of the side of the belt, measured from centre-line to centre-line of the steps at their perimeters, will vary as the sizes of the steps vary. Thus, if we apply a pair of compasses at K L and at M N, we shall find the distance K L the shorter, and as a result the belt would have less tension when on those steps, and the degree to which this will exist will be in proportion to the length of the belt and the variation in the sizes of the steps of the cone.

To remedy this defect, the cones are laid off, as shown in Fig. 349, *a b* representing the axial line of the cone;

Fig. 349.

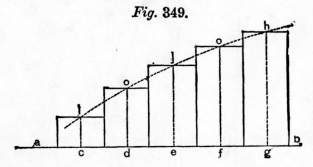

c, d, e, f, g are the centre-lines of the steps; and *h, i* a curve, whose radius will vary according to the conditions. The rule for finding it is, subtract twice the distance between the centres of the two cones from the total length of the belt (measured over the end steps) and divide the remainder by 6·2832, the quotient is the radius of curve required. The process is to mark from the centre-line *a, b* (Fig. 349) the radius of the end steps of the cone, as at *h, i*, to find the required centre of curve from *h, i*, and the intersection of the curve with the centre-lines of the steps, as at *o, o, j*, is the radius for the intermediate steps of the cone.

CHAPTER XIX.

MACHINE TOOLS.

WHAT is known as a machine tool is a machine that operates a cutting tool for dressing the work. *The lathe is the chief of all machine tools,* but the lathe, pure and simple as our forefathers knew it, has of late years undergone many improvements and transformations, widening its scope and very greatly extending and expediting its operations. Thus the screw machine is but a form of lathe designed to produce certain classes of work rapidly and with a minimum amount of skill in the operator.

The shaping machine is being rapidly superseded by the milling machine, which is also trenching upon the field of operations of the planing machine, more especially, however, in the smaller sizes. This occurs not only because milling can be more rapidly executed than planing, but also for the reason that when once the cutters are properly selected and arranged any number of pieces may be milled to one exact size and shape without requiring the operator to measure the work.

Referring now to the simplest form of the lathe.

Fig. 350 represents a foot lathe. It consists of a bed or shears, carrying a head stock for driving the work and a tailstock or footstock for supporting the work at the other end. The headstock is secured firmly to the shears, while the tailstock is made movable along it so that it may be moved up until the dead centre meets the end of the work which is supported at the driving end by the live centre.

The live centre is sometimes dispensed with and a chuck used instead to drive the work, or the work may be clamped or bolted to the face plate. In either of these cases the use of the tailstock and dead centre is dispensed with, unless in the rare case of the work

Fig. 350.

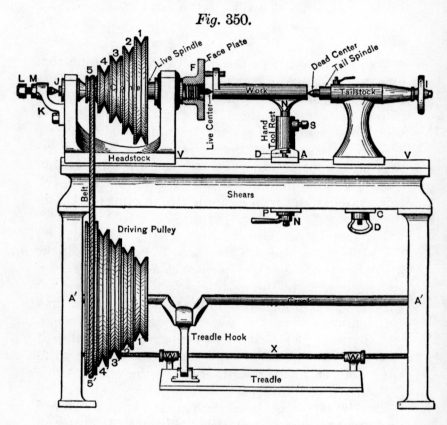

standing some distance out from the chuck or face plate. In the figure the lathe is shown with a hand rest, which is a device whereon to support such cutting tools as are used by hand. The upper piece N, whereon the hand tools are rested, is adjustable in height by means of the set screw S. The body of the hand-tool rest is movable in a direction across the lathe shears to suit the diameter of the work by the following construction: It has a T groove C, receiving the head of a bolt whose threaded end is seen beneath the plate P, which, in conjunction with the two plates A and P, grips the shears when the handle N beneath P is screwed up. The hand nut D is for securing the tailstock or back head, as it is sometimes called, to the lathe shears. The end J of the live spindle, as the shaft or spindle that drives the work is called, is prevented from having any motion endwise by the screw I, which is threaded through the bracket K and has a nut M, to hold it in its adjusted position and prevent it from loosening back, as nuts under some circumstances are apt to do. The driving mechanism is so apparent from the cut as to require no explanation. Lathes of this kind are but very little used because of the amount of labor and skill required in their operation, if good work is to be procured.

We may, however, by providing such a lathe with a slide-rest greatly add to its efficiency, since the slide-rest will not only take heavier cuts than a hand tool but will guide the tool accurately in the required direction.

The construction of a slide-rest may be understood from Fig. 351. The cutting tool, T, is carried, it is seen, in the upper slider, which, by means of a screw whose handle is shown at C, may be moved in or out on the upper slide so as to regulate the depth of cut taken off the work and thus regulate its diameter. To carry the cut along the work the handle A operates a screw whose nut is affixed to the lower slider, hence operating A moves or feeds the lower slider along the lower slide and thus traverses the tool to its cut along the work.

Slide-rests of this class are objectionable and inconvenient for two principal reasons, first, because the handle A comes in the way of the lathe tailstock which will not permit A to pass. This necessitates one of two things, either the handle must be taken off and replaced every time the handle comes against the tailstock, or else the tool must be placed far enough out from the tool-post to permit A to pass. This is a bad element, because for good work the tool should be supported by the slide-rest as near to the work as is conveniently possible.

In the next place the left hand end of the lower slide is apt to come into the way of the work driver.

Fig. 351.

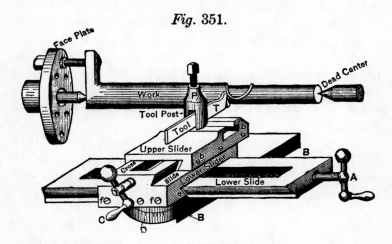

The first of these objectionable features is largely obviated by the construction of a rest such as is shown in Fig. 352, in which the feed screw is located outside the lower slide, as seen in the cut, which represents Warner and Swasey's slide-rest.

Fig. 353 represents a foot-power lathe by W. F. & John Barnes Co., of Rockford, Illinois. There are four changes of speed, and a screw and a rod feed are provided, the latter being for ordinary tool feeding and the screw for screw cutting. The construction of the lathe is so clearly seen in the engraving that but little descrip-

Fig. 352.

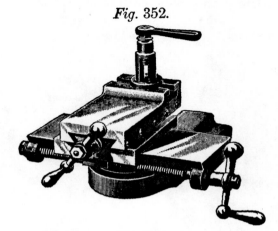

Fig. 353.

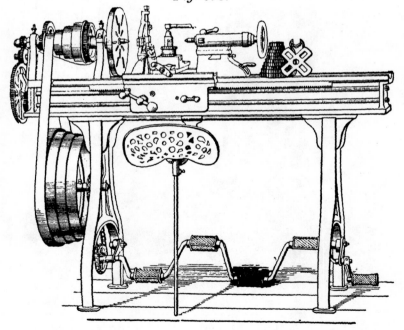

tion is necessary. This is a plain, simple engine

lathe, which means that it is provided with an automatic tool feed and screw-cutting change gears.

Fig. 354.

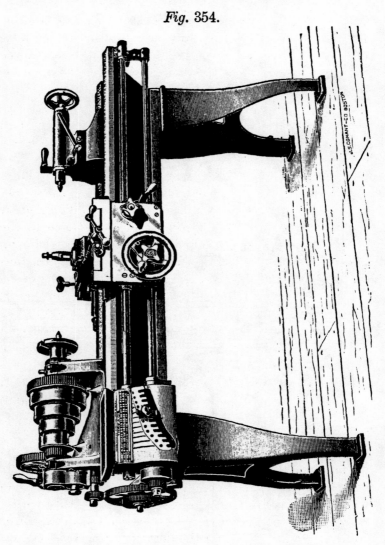

A class of lathe designed to have wide range of tool feed and to cut a large number of threads without requiring to put on or take off any change gears or to move a feed belt is shown in Figs. 354, 355 and 356, which represent the Hendey-Norton lathe.

The gears necessary to cut threads having a pitch from six up to twenty per inch are mounted in the form of a cone directly on the lead screw and are secured thereto by one spline, or key, the whole being enclosed in a case or box, which at once forms the cover for the gears, and the bearings at either end for the screw. In the lower part of this box is arranged a driving shaft with bearings parallel to the screw. This shaft has a spline the full length of the inner side of the box, and has sliding upon it the driving gear, or, as it is commonly termed, the stud gear. This gear bears the proper relation to all the gears in the cone, to cut the regular list of threads from 6 to 20, its position relative to the gears in the cone being controlled by the handle shown, the inner end of which is a forked casting with bearings on either side of the gear, and in an upper extension of the same fork is the bearings for an intermediate gear, which is thrown in or out of the various gears of the cone by means of the handle as shown. The index plate on the front has notches of sufficient depth to receive and guide the handle and gear in perfect line with the cone gear wanted, the thread which the combination will cut being stamped above each notch. The latch for holding the handle and gear in place is arranged to secure the handle, both in and out, entering the upper hole when in, and the upper part of the notch for handle when out. This prevents any possibility of the handle being thrown out from the motion of the shaft or gears when running, and also holds the handle in position where last used, which would otherwise fall to the lower end of the slot. Thus far the device is described as only cutting the 12 regular threads from 6 to 20 (which include all the ordinary threads in daily use), and is accomplished without change aside from the movement of the lever from one notch to the other.

The lower shaft having the same rotation as the lathe spindle, by means of equal gears on the outer end of the shaft, and regular stud of the lathe, it will be seen that changing the relation of these gears will multiply the list of threads according to the ratio of the gears in use,

Fig. 355.

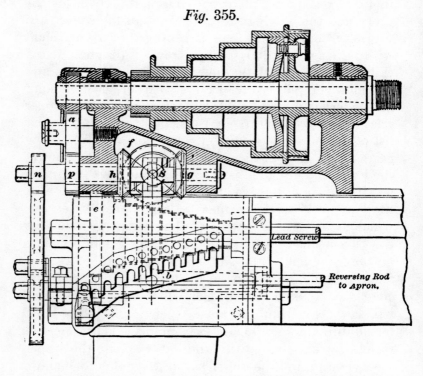

the index having as many rows of figures as there are changes.

The lathe here shown has but two changes, cutting from 1½ threads per inch to 80 threads per inch, and has one extra gear to cut 11½ threads per inch, for steam pipe (which is so often called for), yet, should occasion arise to cut any special thread not provided for, this arrangement does not interfere with making and using any special gear, the same as in any ordinary lathe.

For feed, in turning, this device has the same advantages as for screw cutting, giving 36 distinct feeds, with only two changes of gear.

In daily use there are no changes of gear required for the feeds, as it makes from 30 cuts per inch to 100,

Fig. 356.

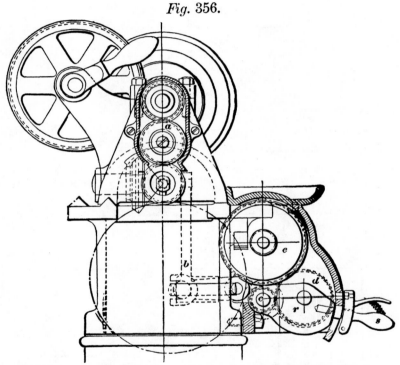

with only the movements of the lever, and with the two changes of gear referred to above will make from 7½ cuts per inch to 400.

Referring now to Figs. 355 and 356, a receives motion from the gear on the live spindle and imparts it through a pinion p, to the bevel gear h, which drives bevel gears f and g. There is a clutch between the two gears, h and

420 COMPLETE PRACTICAL MACHINIST.

Fig. 357.

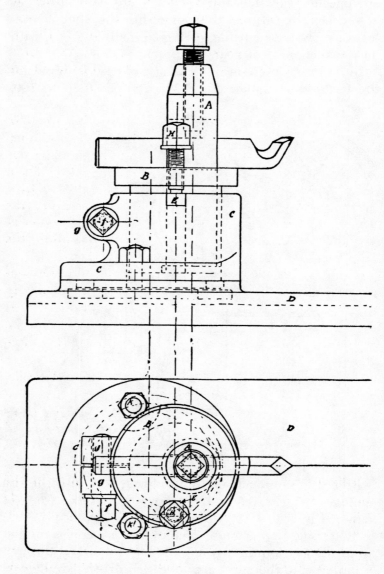

g, operated by an arm, b, which is actuated by partially

revolving the rod marked "Reversing rod" to apron. When the clutch s is moved to engage with gear g the gear n is revolved in one direction, while, when s is moved to engage with the gear h, wheel n is revolved in the opposite direction. Now since n operates the lead screw gears it is obvious that their direction of revolution will be reversed whenever clutch s is moved from meshing into gear h to gear g, or vice versa.

The construction of the tool post is seen in Fig. 357. A is secured in c in the usual manner. B is cylindrical and fits into c, which is split through, and has two lugs g which may be gripped by tightening the bolt or screw f or released by loosening it. Now suppose f to be tightened slightly, or left loose, and by operating the screw H, which is threaded through B and rests upon E, the tool post A can be raised or lowered to adjust the height of the tool to the work.

Fig. 358 represents a screw-cutting engine lathe constructed by the New Haven Manufacturing Company of New Haven, Conn. The rod feed motion of this lathe is so constructed that it may be, by a single motion, set to either of three rates of feed, while at the same time the direction of the feed may be reversed.

The spindle of this lathe is hollow, so that the rods, etc., may be passed through it.

The compound rest is provided with a self-acting, or automatic, cross feed, and the screw feed and rod feed motions are so arranged that they cannot be both put into action at the same time.

The bed or shears is, it will be seen, deepest in the middle to avoid deflection, which occurs to a sensible degree in all lathes whose shears are light in proportion to the work the lathe may be called upon to do.

A monitor lathe is one in which the cutting tools are carried in a turret that is capable of revolving upon its centre and in a horizontal plane.

Fig. 358.

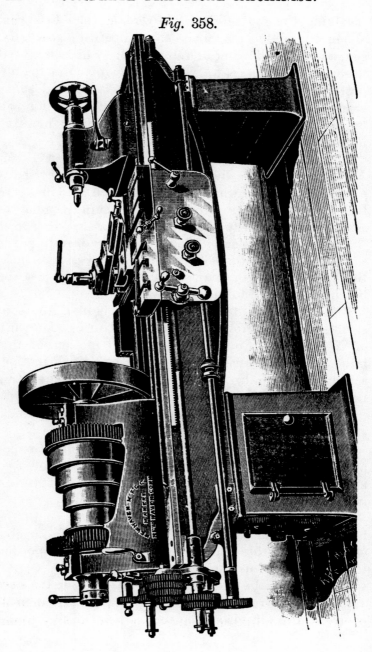

MACHINE TOOLS. 423

The turret is usually provided with six holes, so that six tools may be used. It is so constructed that the

Fig. 359.

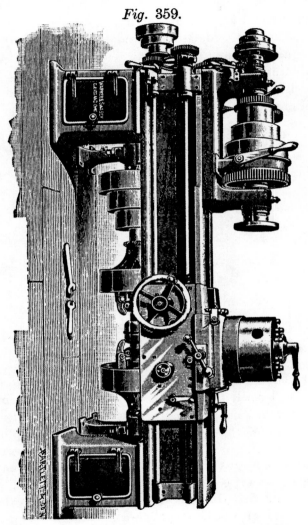

movement of the turret occurs automatically at the end of the back traverse of the turret, bringing the tool exactly in the position required for it to operate upon the work on its forward traverse, which may be done by hand or by an automatic feed. The diameter to which the work will be cut is determined by the position in which the tool is set in its holder.

Fig. 359 represents Warner and Swasey's twenty-four inch universal monitor lathe. The head is provided with a three-stepped cone and with back gear. The turret is provided with an automatic cross feed in both directions, and the carriage with a screw-cutting feed and an ordinary turning feed. This lathe is provided with a friction clutch, by means of which the speed

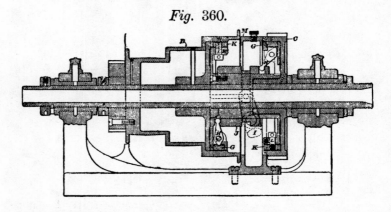

Fig. 360.

may be changed from the ordinary cone speed to back gear speed while the lathe is in motion; and as many modern lathes are provided with this convenient feature an example of the means by which it is accomplished is given in Figs. 360 and 361. By a simple movement of the lever A, at the side of the cone, the speeds can be varied in the ratio of four to one, thus giving the proper changes for the different diameters in the piece being turned, and also between the boring and tapping. Its construction can be readily followed by reference to the figures, which represent a longitudinal section through the head and also a cross section through the clutch mechanism. In this form of head both the cone B and the head gear C run loose on the spindle. Between the two is the driver D keyed fast to the spindle. This driver consists of a hub carrying a head at each end, the one at the gear end being cast on, while the one at the cone is screwed on, to admit of assembling the parts.

In the periphery of each of these heads is turned a rectangular groove, and in these grooves fit rings E, E. These rings are made in halves, each half being fastened at one end to the head by the screws F. The other ends of the rings are beveled, and between them fits the wedge G.

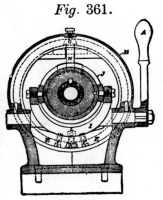

Fig. 361.

Motion is imparted to this wedge by the toggle H, and this in turn is actuated by the lever A, through the fork I and sleeve J. Thus, when the lever A is thrown towards the head gear the wedge in the head gear ring is thrown out, the ring expanded and the gear is clutched to the spindle. When the lever is thrown toward the cone, the head gear is first released, and then the wedge in the cone ring is thrown out, the ring expanded and the cone clutched, as in the case of the head gear. When the lever is vertical both the head gear and the cone are free, and the spindle remains stationary.

An excellent feature of this clutch is the means provided for taking up wear and adjusting the rings. Between the fixed ends of the rings E, E, and opposite the actuating wedge G is a second wedge K, similar to the other, except that it is reversed, the small end being toward the centre. Into this wedge is tapped the adjusting screw L, which can be easily worked from the outside when the shield M is removed. The wedge K also serves another purpose. It is made long enough to pass through the ring and across the flanges of the head on either side. The ends of the wedge where it passes through the flanges being squared, it thus serves as a key or driver for the rings, taking all side strain from the screws.

Fig. 362 represents a lathe in which the spindle for the dead-centre is square and the tail-stock is capable of being moved or fed crosswise of the lathe, after the manner of a slide rest. In this class of lathe boring or cutting tools may take the place of the dead-centre and

Fig. 362.

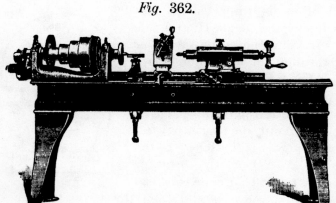

the tail-stock be operated as a slide rest; this being mainly suitable for brass work.

Turret lathes are now being made in much larger sizes than formerly, and so much solidity being put into the parts that they are enabled to take very heavy cuts on large work, thus greatly expediting its execution and placing it far beyond the ordinary form of lathe.

Fig. 363 represents a turret lathe of this kind manufactured by the Gisholt Machine Company of Madison, Wisconsin. In addition to the ordinary turret the lathe is provided with four tool posts, shown at *e*, so that

MACHINE TOOLS.

Fig. 363.

including those carried in the turret, ten tools can be set

to operate upon the work, any two of which can be cutting at the same time. This is a triple-geared lathe, the third motion gearing direct into the rim of the face plate. The feed screw has a quadruple thread and the result is that when the nut is released the screw will, from its coarse pitch, revolve the nut; the turret carriage ceasing its automatic feed until the nut is held stationary, which is done by a clutch motion operated by the lever C. The arm A operates the stop motion, a small arm at f engaging with the rod or shaft that carries the stops, which are seen at b. There is a separate stop for each tool held in the turret, and each, of course, may be set to meet the requirements of the tool whose amount of motion it is required to govern.

What is known as a screw machine is a special form of lathe in which the work is cut direct from the bar, without the intervention of forging operations, and it follows therefore that the bar must be large enough in diameter to suit the largest diameter of the work, the steps or sections of smaller diameter being turned down from the full size of the bar. The advantages of the screw machine are, that the work requires no centring since it is held in a chuck, that forging operations are dispensed with, that any number of pieces may be made of uniform dimensions without any measuring operations save those necessary when adjusting the tool for the first piece, and that it does not require skilled labor to operate the machine after the tools are once set, which is usually done by a specially skilled workman.

The capacity of the screw machine is, therefore, many times greater than that of a lathe, while the diameters and lengths of the various parts of the work will be more uniform than can be done by caliper measurements, being in this case varied by the wear of the cutting edges of the tools only, which eliminates the errors liable to independent caliper measurement. Hollow

MACHINE TOOLS.

work, as nuts and washers, may be equally operated on being driven by a mandril held in the chuck.

Fig. 364 represents a small screw machine.

Three separate tool-holding devices may be employed: first, cutting tools may be placed in the holes shown to pierce (horizontally) the circular head F; second, tools may be fixed in the tool posts shown in the double slide rest, which has two slides (one in the front and one at the back of the line of centres); and

Fig. 364.

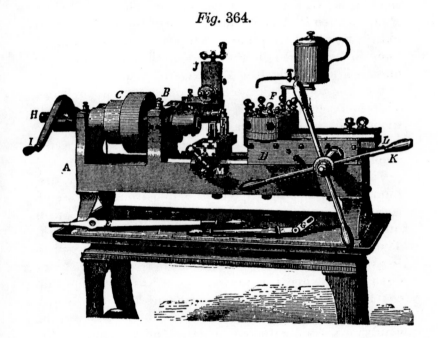

third, tools may be placed in what may be termed the screw-cutting slide rest J.

F is a head pierced horizontally with seven holes, and is capable of rotation upon L; when certain mechanism is operated L slides on D and the mechanism of these three parts is arranged to operate as follows: The lever arms K traverse L in D. When K is operated from right to left, L advances towards the live spindle until arrested at some particular point by a suitable stop motion, this stop motion being capable of adjustment so as to allow F to approach the live spindle a distance suitable for the work in hand.

When, however, K is operated from left to right L moves back, and when it has traversed a certain distance, the head F rotates $\frac{1}{7}$ of a rotation, and becomes again locked so far as rotation is concerned. Now the relation between the seven holes in F is such that when F has rotated its $\frac{1}{7}$ of a rotation, one of the seven holes is in line with the live spindle. Suppose then seven cutting tools to be secured in the holes in F, then K may be operated from right to left, traversing L and F forward, and one of the cutting tools will operate upon the work until L meets the stop; K may then be moved from left to right, L and F will traverse back, then F will rotate $\frac{1}{7}$ of a rotation and L and F may be traversed by K, and a second tool will operate upon the work, and so on.

The diameter of the work is determined by the distance of the cutting edge of the tool from the line of centres, when such tool is in line with the work, or, in other words, is in position to operate upon the work. The end measurements of the work are secured by placing the cutting edges of the tools the requisite distance out from F, when L is moved forward as far as the stop motion will permit. But it is evident that the length of cut taken along the work would, under these simple conditions, vary with the distance of the end of the work from the face of the chuck driving it, but this is obviated as follows:

The live spindle is made hollow so that the rod of metal, of which the work is to be made, may pass through that spindle. A chuck on the spindle holds the work or releases it in the usual manner. Suppose then the chuck to be open and the bar free to be moved, then there is placed in the hole in F, that is in line with the work, a stop instead of a cutting tool. The end of the work may then, for the first piece turned, be squared up by a tool placed in the slide rest and then released from the chuck and pushed through the live spindle until it abuts against the stop so adjusted and affixed in the hole in F. K may then be operated to act on the work. The first tool may reduce the work to its largest required diameter, the second turn down a plain shoulder, the third may be a die cutting a thread a certain distance up the work, the fourth may be a tool turning a plain part at the beginning of the thread, the fifth may round off the end of the work, and the sixth may be a drill to pierce a hole a certain distance up the end of the work.

Now suppose the work to require its edge at the other end to be chamfered, then there may be placed in the slide rest tool posts a tool to sever the work from the bar out of which it has been made, while the other may be used to chamfer the required edge, or to round it if need be to any required form.

Work held in the chuck but not formed from a rod may be, of course, operated upon in a similar manner.

In the case, however, of work of large diameter requiring to be threaded, the threading tool may be held and operated differently and more rigidly, as follows: I is a lever carrying under its bend, and over the projecting end of the live spindle, a segment of a nut whose thread must equal in pitch the pitch of thread to be given to the work. A collar or ring, oftentimes called the leader, having a thread of the same pitch, is then secured upon the live spindle, so as to rotate with it, and have no end motion; when therefore I is depressed, the nut will come into work with the leader or ring, and I will be traversed at a speed proportioned to the pitch of the threads on the collar and nut.

Now I is attached to a shaft having journal bearing (and capable of end motion) at the back of the lathe head, and on this bar is attached the slide rest J, in which the turning or threading tool may be placed. The shaft above referred to, having end motion, may be operated (when the nut in the lever I is lifted clear of the collar) laterally by means of the lever I; hence to traverse J to the right, or for the back traverse, I is raised and pulled to the right, I is then lowered, the nut engages with the collar, and the tool is traversed to the cut. The cut is adjusted for diameter by the slide rest, which is provided with an adjustable stop to determine the depth to which the tool shall enter the work.

It is obvious that this part of the machine may be employed for ordinary turning operations, if the collar be of suitable pitch for the feed.

Fig. 365 represents the Brown & Sharpe Manufacturing Co.'s screw machine, which is designed for making studs, etc., from the bar and for finishing work held in a chuck when several tools require to be used or a number of operations performed. The back gears are under the spindle and are enclosed. They run continuously and are engaged or disengaged by a clutch motion. There are eight changes of automatic turret-slide feed. The feed cones have four steps, and by shifting a lever each of the four speeds can be made fast or slow without changing the feed belt.

The cutting-off machine is a form of lathe used to cut rods or bars into exact lengths, which saves a great deal of time over the old plan of cutting them on the anvil at the blacksmith's fire. In the first place, the bar can be cut in less time, and in the next, the cutting-off machine leaves the end of the work square and true and of the required length.

The spindle of a cutting-off machine is hollow and

MACHINE TOOLS. 433

the bar passes through it until it comes up to a stop or gauge which determines the distance the bar shall pro-

Fig. 365.

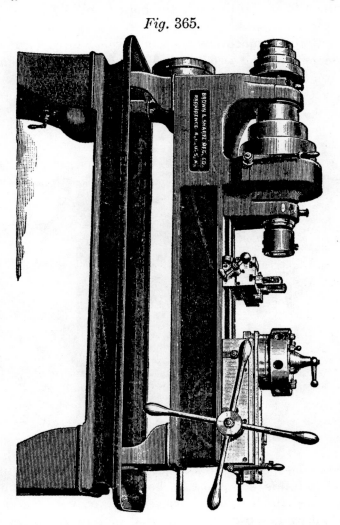

ject beyond the cutting tool, this distance obviously being the length of the piece cut off. Chucks at each end of the live spindle are provided to hold and to guide the work, the driving chuck to open to release the work and close to grip it automatically, and one or the other of them to feed it through the hollow spindle up to the length gauge.

Cutting-off machines usually have a simple form of slide rest having two tool-holding heads on the same

Fig. 366.

cross-slide, the tool on the second or back head being turned upside down so that both tools cut at one and the same time. This, however, can only be accomplished by so setting the tools that each meets the work at the same time as the other, in which case each tool takes an equal depth of cut and the tool feed can be twice as fast as it would be if only one tool were used.

The cutting-off machine is not now so important as it was because a good many turret lathes are constructed to perform the duties of a cutting-off machine even when the work is long and cut from full-length bars.

Fig. 366 represents a cutting-off machine constructed by the Hurlburt Rogers Machine Company. This machine possesses a valuable feature in that the speed of the work revolution increases as the cutting-off tools approach the centre of the work, so that a constant rate of tool-cutting speed is attained, and as a result the work is greatly expedited. The form of tool used in this machine is a parallel blade of the required width, ground accurately to shape and with clearance given on the side of the tool, so that all the operator requires to do to resharpen the tools is to grind them on the end face, not touching either the side faces or the top face. The front rake necessary to make the tool cut keenly and well is given by so setting the tool that the top face instead of lying in an horizontal position lies at an angle of about twenty to thirty degrees, the cutting edge, of course, being the highest and coming level with the centre of revolution of the work.

Shapers and planers are machines in which the tools cut in a straight line. In a shaper the tool is moved across the work, whereas in a planer the work moves and the tool is held stationary. An example of a shaping machine is given in Fig. 367, in which a column or frame has a shaft running through it, carrying at one end a belt pulley P, and this shaft drives a pinion p, which in turn drives a gear wheel Q, on a shaft, which at its other end drives the rod S, which, being secured to the ram or bar A, causes it to traverse to and fro endwise. To cause A to traverse in a straight line it is fitted into a slideway provided in the top of the frame.

The cutting tool T is carried on the end of the ram and the work is held either in a vice, shown at V, V^1, or some other work-holding chuck or device, which is bolted or clamped to the work-table W.

The tool-carrying head J is so connected at G to the ram or bar that it can be swung in a vertical plane, and thus set the cutting tool to stand out of the vertical when the conditions require it. The feed screw, whose

Fig. 367.

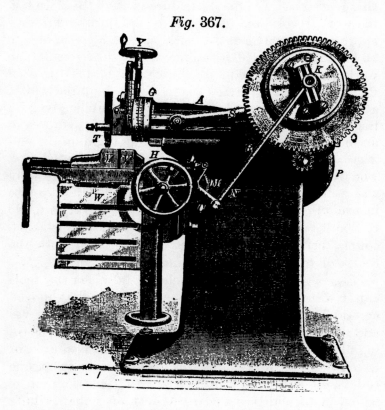

handle is shown at Y, moves the tool carrying slider in a straight line, which may be either vertical or otherwise, according to the requirements.

A front view of the tool-carrying head is shown in Fig. 368, in which Y is the handle for the feed screw operating the slider I, upon which is the apron swivel

J, in which is pivoted the apron K' carrying the cutting tool. Apron K' is pivoted by a taper pin K, so that the tool can be lifted upon the return or back stroke, which prevents the tool edge from rubbing on the back stroke and thus getting dulled.

When the tool is carrying a cut in a vertical direction the apron swivel J is set over, as shown in the figure, and in that case lifting the tool on the back or non-cutting stroke causes its cutting edge to be well clear of the work, which is obviously a great advantage. To cause the stroke of the ram or bar A to traverse across the work to no greater distance than the length or width of the work, requires the rod S, driven by a pin which may be set to stand at an adjusted distance from the shaft of the gear wheel K, Fig. 367, it being obvious that the nearer this pin is set to the centre of its driving shaft the shorter the stroke of the ram A will be.

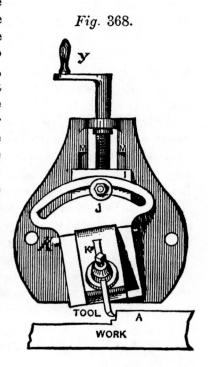

Fig. 368.

This part of the mechanism of a shaper is usually so constructed as to cause the tool to move faster on its back return or non-cutting stroke than it does on its cutting stroke, which obviously expedites the work.

Now suppose the work to be set in position between the vice jaws V V', and the tool to be set to the required depth of cut, and the machine will feed the work table and therefore the work, so that the cut shall be carried across it by means of the following construction—referring to Fig. 367. A slotted piece K is secured to the gear wheel Q, and a pin in K drives the rod R, which vibrates the arm C of the bell crank, which has at D a catch, which for feeding is caused to engage with the teeth of the gear wheel M, which engages with a pinion on the shaft of the feed traverse screw, which operates the work table W in a direction at a right angle to the line of motion of the ram or bar A.

It is obvious that at each revolution of the driving gear Q the rod R vibrates the bell crank once back and once forth. Now suppose the arm C of the bell crank to move forth to the right, and its upper arm will move to the left, and the latch seen at D will engage with the teeth of the gear wheel M, and rotate it through a part of a revolution commensurate with the amount of motion of the bell crank.

When the bell crank reverses its direction of motion, the catch on its upper arm, being pivoted, will lift and pass over the teeth of the gear wheel M, leaving it stationary; thus the feed or traverse of the work-table is intermittent, one feed of the work-table (and therefore of the work) occurring to each cutting and return stroke of the cutting tool. This feed can be caused to begin at either the beginning of the return stroke or at the beginning of the cutting stroke, as may be desired, by altering the position of the pin in K, Fig. 367, it being obvious that when that pin is located on one side of the centre of the gear wheel Q, its motion will be the opposite to what it would be if it were placed on the other side of that centre. Take the position the parts occupy in the figure, and it is obvious that if the top of the piece K is moving to the right, then the bottom of K will be moving to the left. If the operator lifts the tool on the back or non-cutting stroke, it does not

Fig. 369.

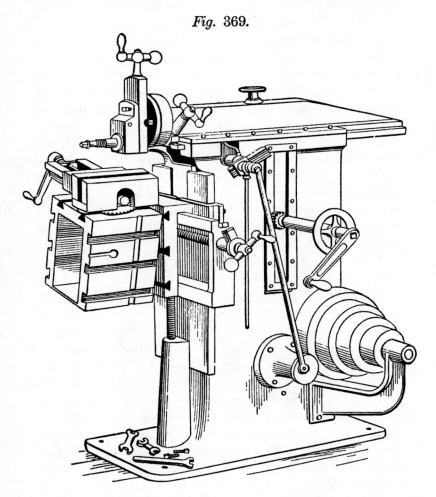

matter on which stroke the feed occurs; but if he does not, the tool will rub against the cut on the return stroke, no matter on which stroke the feed occurs.

Fig. 369 represents Fox's pillar-shaping machine, in which the stroke of the ram may be adjusted while the machine is in full motion, which is accomplished as follows: Within the frame of the machine is a slotted bar or link, which at the top is pivoted to the ram of the machine. At the lower end of the slotted link there is fitted into the link slot a die upon a crank pin, provided in a wheel, and it is obvious that as the crank pin revolves the die moves up and down in the link slot, while at the same time the lower end of the link vibrates back and forth, the upper end of the link vibrating the ram back and forth for the cutting and return strokes within the slot of the links, and somewhere between the pivoted top and the crank pin and die is a slide block acting as a centre, on which the link vibrates, and it depends upon the height of this die what the length of the tool stroke shall be. The higher the slide block is, the less the ram stroke, and vice versa. The height of the slide block is regulated by the rack and pinion shown at the side of the machine, which may be adjusted while the machine is in full motion.

In some classes of shapers (called traverse shapers) the work is held stationary and the tool-carrying ram or bar is traversed in a slide or saddle, as seen in Fig. 370, in which A is the ram setting on the slide B, which is capable of being traversed or fed along the slideway provided on the upper surface of the bed, M, M, of the machine, by means of a screw, the feed wheel and catch for which is seen at F. At G is a worm engaging with teeth arranged on the arc of a circle on the top of the tool head, in order to swivel that head and cause the tool to clear the cut on the return stroke.

The two pieces D are adjustable anywhere along the

MACHINE TOOLS. 441

Fig. 370.

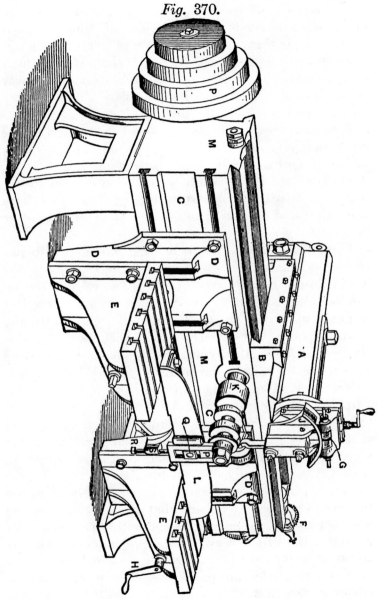

front face C of the machine, their office being to support the work tables E E, whose height may be regulated by means of handles, one of which is seen at H, which through the medium of gear wheels operate vertical screws, which, being threaded through nuts provided to E E, cause them to raise or lower as may be necessary to bring the work to the proper height beneath the cutting tool.

It is obvious that if the work is long enough to require it, one end may be set upon one of the tables E, and the other end upon the other table.

The device shown at K is called a cone-mandril, which is used to hold work having a hole or bore through it and requiring to be dressed or cut to an arc of a circle.

This device consists of an arbor or mandril with two cones upon it, one of these cones being fast upon the mandril and the other removable. The work is put on the mandril and the loose cone is then put on and screwed up (by a nut on the mandril end) until the work is firmly held between the two cones. The cutting tool is fed vertically by hand to set the cut, and the mandril is, by suitable mechanism provided in the machine, revolved through a portion of a revolution at each cutting stroke or traverse of the ram.

The beam shown at L is a support for the outer end of the cone mandril, the support being given by the piece P, whose height is adjusted by the set screw R, which is threaded through the lug S.

Fig. 371 represents a pillar shaper by the Hendey Machine Company. This is what is sometimes termed a planer shaper, because the ram motion is reversed by reversing the belt motion, there being two belt pulleys running in opposite directions. Of the two dogs seen beside the ram, the left-hand one trips a lever which causes a friction clutch to grip the pulley that drives the ram on the cutting stroke, while the right-hand one causes the clutch to release its grip on the forward pulley and transfer the grip to the quicker running pulley that drives the ram on the return stroke, and so accurate is the mechanism in its operation that a tool stroke of not more than $\frac{1}{16}$ inch can be taken, the cut ending if

Fig. 371.

needs be against a shoulder or projection without breaking the tool point. The length of stroke is obviously determined by the distance apart at which the dogs are set, and they can be moved and adjusted with perfect ease while the machine is in full motion.

Morton's shaping machine, having some novel and interesting features, is shown in Fig. 372. In this machine the tool cuts on the back stroke and is therefore pulled to its cut instead of being pushed to it, as is common to other shaping machines. This enables very heavy cuts indeed to be taken (that is to say, cuts as much as an

Fig. 372.

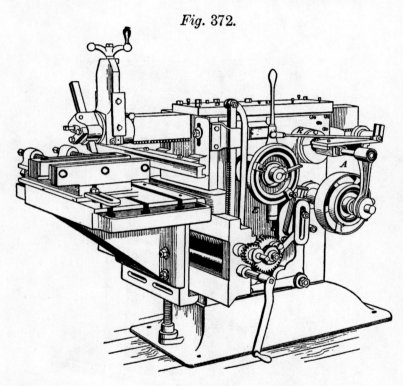

inch and a half deep in cast-iron), which could not be done in a machine in which the tool cuts on the forward or outward stroke. The ram or bar for carrying the tool head (which has the usual vertical hand-feed motion) is, it will be seen, square, and has a rack running along and secured to its underneath or bottom face or surface.

This rack is driven by a spur gear receiving motion from a train of gears driven from the belt pulley A. The smaller pulley, shown on the right of A, is operated by a separate belt and is for the quick return motion of the tool.

These two pulleys are provided with a friction device so constructed that either of them may be allowed to run idle without operating the machine, or either may be locked to the driving shaft so as to drive the machine, and it is by alternate locking one or the other that the ram is operated back and forth as required.

The operating mechanism may be thus briefly described: A gear wheel within the frame of the machine operates a friction disk, E, to which is attached the handle F, operating the rod r, and it is this rod r that operates the clutches of the driving pulleys. It will be seen in the engraving that the rack beneath the ram is almost in direct line with the edge of the cutting tool, an element that adds greatly to the rigidity of the machine.

The iron planing machine, or iron planer, as it is termed in the United States, is employed to plane such surfaces as may be operated upon by traversing a work table back and forth in a straight line beneath the cutting tool. It consists essentially of a frame or bed A, Fig. 373, provided on its upper surface with guide ways, on which a work carrying table T may be moved by suitable mechanism back and forth in a straight line.

This frame or bed carries two upright frames or stanchions B, which support a cross-bar or slide C, to which is fitted a head which carries the cutting tool.

To enable the setting of the tool at such a height from the table as the height of the work may require, the cross slide C may be raised higher upon the uprights B by means of the bevel gears F, G, H and T, the latter being on a shaft at the top of the machine, and operating the former, which are on vertical screws N, which pass down through nuts that are fast upon the cross slide C.

To secure C at its adjusted height, the uprights are provided with T-shaped slots H, H, and bolts pass through C, their heads being in the T-grooves, and

Fig. 373.

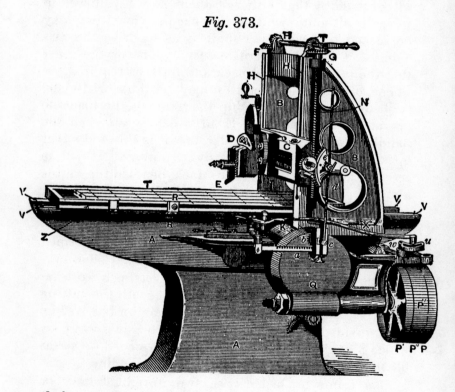

their nuts exposed so that a wrench may be applied to them.

The faces of the cross slide C are parallel one to the other, and stand at a right angle to the V-guideways on which the work table (or platen, as it is sometimes termed) slides; hence the cross slide will, if the table is planed true or parallel with this cross slide, be parallel with the table at whatever height above the table it is set, providing that the elevating-screws, when operated, lift each end of C equally.

The construction of the head D corresponds to that of the head shown in Figs. 367 and 368 for a shaper, except that in this case the swivel head is secured to a saddle that slides along C, being provided with a nut operated by a feed screw J, which moves D along C.

The mechanism for operating the work table or platen T is as follows: P P' are two loose pulleys, and P'' is a driving pulley fast on the same shaft. This shaft passes beneath the bed and drives a worm operating a wormwheel, which actuates inside the frame A and beneath the work table a train of gears, the last of which gears with a rack, provided on the underneath side of the table.

The revolutions of this last wheel obviously cause the work table to slide back and forth while resting on the V-guideways provided on top of the frame A, the direction of table motion being governed by the direction in which the driving pulley P'' revolves.

This direction is periodically reversed as follows: The pulley P is driven by a crossed belt, while pulley P' is driven by an open or uncrossed one; hence the direction of revolution of the driving pulley P'' will be in one direction if the belt is moved from P to P'', and in the other if the belt is moved from P' to P''. Mechanism is provided whereby first one and then the other of these belts is moved so as to pass over upon P'' and drive it, the construction being as follows:

To the edge of the work table there is fixed a stop R, which as the table traverses to the right meets and moves a lever arm S, which through the medium of a second lever operates the rod X, which operates a lever *u*, which has a slot through which one of the driving belts passes. The lever *u* operates a second lever *w* on the other side of the pulleys, and this lever also has a slot through which the other driving belt passes.

When the stop R moves the lever arm S, levers u and w therefore move their respective belts, one moving from the tight pulley P″ to a loose one as P, and the other moving its belt from the loose pulley as P′ to the tight one P‴, and as the directions of belt motions are opposite, the direction of revolution of P‴ is reversed by the change of belt operating it. There are two of the stops R, one on each side of the lever S; hence one of these stops moves the lever S from left to right and the other from right to left.

Suppose, then, that the table is moving from right to left, which is its cutting stroke, and the driving belt will be on the pulley P″ while the other belt will be on pulley P. Then as the stop R moves S and operates X the arm u will move its belt from P″ to P′, and arm w will move its belt from P to P‴, reversing the direction of motion of P‴, and therefore causing the table T to move from left to right, which it will continue to do until the other stop corresponding to R meets S and moves it from right to left, when the belts will be shifted back again. The stroke of the table, therefore, is determined by the distance apart of the stops R, and these may be adjusted as follows:

They are carried by bolts whose heads fit in a dovetail groove Z provided along the edge of the table, and by loosening a set screw may therefore be moved to any required location along the bed.

To give the table a quick return so that less time may be occupied for the non-cutting stroke, all that is necessary is to make the countershaft pulley that operates during the back traverse of larger diameter than that which drives during the cutting traverse of the table.

In order that one belt may have passed completely off the driving pulley P″ before the other moves on it, the lever motions of u and w are so arranged that when the belt is moving from P″ to P lever u moves in advance of lever w, while when the other belt is being moved from P″ to P′ lever w moves in advance of lever u.

To enable the work table to remain at rest, one driving belt must be upon P and the other upon P′, which is the case when the lever arm S is in mid position, and to enable it to be moved to this position it is provided with a handle K forming part of lever S.

To cause the tool to be fed to its cut before it meets the cut, and thus prevent it from rubbing against the side of the cut, the feed takes place when the table motion is reversed from the back or return stroke to the cutting or forward stroke by the following mechanism:

At a is a rack that is operated simultaneously with S and by the same stop R. This rack operates a pinion b which rotates the slotted piece c, in which is a block that operates the vertical rod d, which is attached to a segmental rack e, which in turn operates a pinion which may be placed either upon the cross-feed screw J or upon the rod above it, the latter operating the vertical feed of the tool head. Thus the self-acting feed may be transferred to the cross-feed screw or the vertical feed at pleasure.

Fig. 374 represents a planer of a somewhat larger size, the bed being supported at the ends as well as in the middle. The belt shifting and the feed motion mechanism of this machine are as follows:

The feed motion is operated by a disk C which is actuated one-half a revolution when the work table reverses its motion. This disk is provided on its face with a slideway in which is a sliding block that may be moved towards or away from the centre of C by means of the screw whose head is seen at T, and it is obvious

Fig. 374.

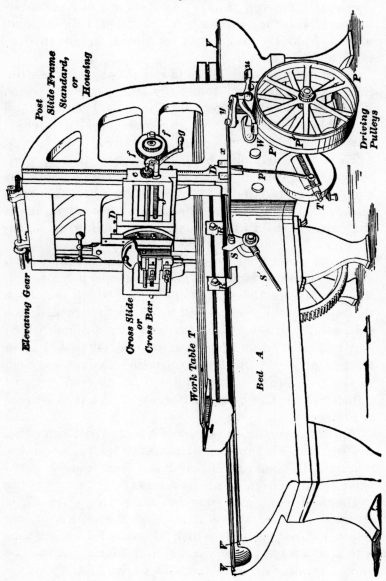

that the further out the block is set from the centre of C the greater the amount of motion that will be given to the rod p, and therefore to the vertical rack D that operates the feed through the medium of the gear wheels shown at f.

The handle g is for operating the feed screw when the automatic feed is put out of action.

The construction of the belt-shifting mechanism is as follows:

R, R' are the actuating stops fastened by their respective nuts at the distance apart necessary for the length of the work to be planed. These stops actuate the lever S (the handle S' being to operate belt shifters by hand when required), S operates the rod x which is attached to a pivoted arm or piece y in which is a cam-slot through which pass pins from the belt moving arms u and W. The shape of the cam-slot is such as causes one of these arms to move a little in advance of the other, and thus to move one of the belts entirely off the centre pulley P before the other belt moves off it, which avoids both strain and noise, this central pulley being that which operates the work table while the other two pulleys (one on each side of P) run idle, one with an open and the other with a crossed belt.

In the Detrick & Harvey Machine Company's open side planer, shown in Fig. 375, there is only one housing or side frame, which enables the machine to do work that would be too wide for a similar-sized machine having two housings or posts.

The machine is as solid and rigid as the ordinary kind of planer and more convenient for a large class of work. The machine has the usual quick return motion and automatic feed motions, while the cross-slide may be lowered or raised by belt power through the medium of the pulleys seen at the top of the machine.

A separate cutting head is provided, as shown on the

side frame, a construction common to large planers, *Fig.* 375.

and frequently there are two separate heads on the cross bar, as seen in Fig. 376.

MACHINE TOOLS.

Milling machines. The milling machine employs rotary cutters to do much of the work that is done by the shaping and planing machine, and to do other work that no other machine will do so efficiently; as, for example, the cutting of spirals and the dressing to shape of complicated forms. A plain milling machine is one in which the work table has simply a longitudinal and a crosswise feed, the latter always moving in a line at a right angle to the spindle which carries the revolving

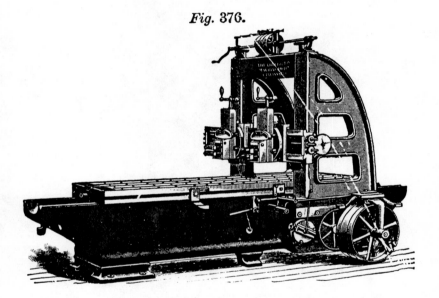

Fig. 376.

cutters. The milling machine is one of the most efficient and valuable of all machine tools on account of the accuracy and rapidity with which it will do a wide range of work.

Fig. 377 represents a Brown & Sharpe Number 0 plain milling machine. It is provided with a four-stepped cone and automatic feed motions both longitudinally and crosswise of the work table. In the figure the handle is shown on the screw for elevating the table.

The middle screw is for traversing the table longitudinally and the upper screw for moving it crosswise; the

Fig. 377.

collar *e* of this screw is provided with graduations, by means of which the cut can be set to the thousandth of an inch, and similar graduations are provided on the elevating screw. The automatic feed is actuated from the feed cones shown, which operate a worm and worm wheel, the latter being on the feed screw of the table.

Fig. 378.

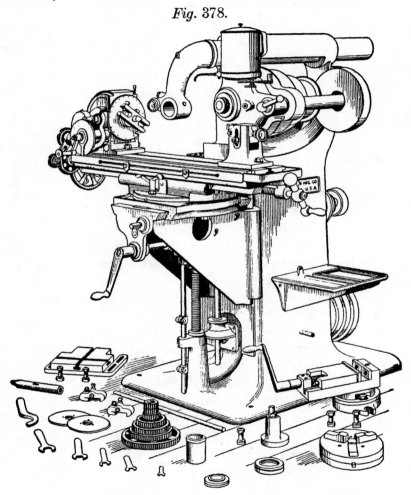

When this table has traversed until its stop *a* meets the belt crank *b* it trips the crank and the worm at *c* becomes disengaged and the feed stops.

Fig. 378 represents a Brown & Sharpe Universal

Milling Machine No. 3. In a universal machine the work table is pivoted so that it can be set to feed the work at an angle other than a right angle to the plane of revolution of the cutters; this being necessary in cutting spirals and other similar work. The saddle which carries the table pivots in the clamp bed and is rigidly clamped to it by three bolts, which slide in circular grooves and allow the table to be set at any required angle up to 45 degrees each way from zero.

The feed of the table is automatic in either direction and can be changed by the simple movement of the lever on the front of the saddle, and as it is driven from the centre it can be used with the table clamped at any required angle to the axis of the spindle. There are twelve changes of feed and the table may be operated by hand from either end.

Adjustable dials on the collars of the feed screws, graduated to read to thousandths of an inch, indicate the longitudinal, vertical and transverse movements of the table, and enable the cut to be set exact. These dials may be operated by hand without the aid of any wrench or screw-driver.

The spiral head has indexing mechanism by which the periphery of a piece of work may be divided into equal parts, and the velocity of the rotary motion of its spindle, or of the work, relative to the speed of the feed screw, is regulated by change gears at the end of the table. Any spiral of the 68 provided for may be cut without interfering with the divisions obtainable from the index plates sent with the machine. A plate for rapid indexing of work into 24 or less divisions is placed directly on the spindle, and the worm which turns the spindle may be thrown quickly out of gear by a knurled knob on the back of the spiral head to allow for this direct or plain indexing. The spindle of the spiral head may be moved continuously, or through any required portion of a revolution; and by use of the raising block the spiral head may be set at any angle on the table. A taper hole 1.05″ diameter at the small end extends through the spindle, and is fitted to receive the same collets and arbors that are used in the main spindle. The front end of this spindle is also threaded, and this end of the spindle may be elevated to any desired degree between 10 degrees below the horizontal and 10 degrees beyond the perpendicular, and rigidly clamped, if desired, at any point. The side of the spiral head is graduated in degrees to show the angle of the elevation of the spindle. The spiral head and foot-stock centres swing 10″ in diameter and take 15″ in length.

The foot-stock spindle may be raised vertically and set at an angle in a vertical plane. By this arrangement the spiral head and foot-stock spindles may, in ordinary use, be kept in line when the front of the spiral head spindle has been elevated or depressed.

The frame is hollow and fitted as a closet to hold the small parts that accompany the machine. On the left side of the frame there is a pan for holding small tools, etc., and on the front of this there is a rack for wrenches. On the other side of the frame is a shelf for holding the spiral head and vise when they are not in use.

Fig. 379 represents a very solid and efficient milling machine of the Brown & Sharpe Manufacturing Company. In this machine the work table has no vertical adjustment but rests solidly on the frame, the height of the cutters from the work being adjusted by the head itself, which is pivoted at one end and seated in and rigidly bolted to a circular seat provided in the solid frame. As a result very heavy cuts and accurate work may be done. The feed table has automatic feed, which can be released at any point and quickly returned or operated by hand, and there are three changes of feed.

Fig. 379.

This machine is suitable where large quantities of duplicate work is done.

Drilling machines, which are often, but improperly,

called drills, are usually made to carry the drills vertically, and are provided in the larger sizes with back gears to increase their power and various hand and power feed motions. Examples of drilling machines are given as follows:

Fig. 380 represents Quint's six-spindle turret drilling machine for drilling and tapping small holes. The bevel gear seen on the cone spindle drives the two bevel gears A and B, which drive within the turret head two friction disks corresponding in form to a pair of bevel gears set face to face, so that a bevel friction pinion or roll can be placed between them and be driven by them. In order to do this it is necessary that the disks revolve in opposite directions, which is done because one disk is driven by the gear A and the other by gear B.

Each of the six drill spindles has its own friction pinion or roll, and the construction is such, only when a drill spindle is in the proper position to operate upon the work does its friction pinion come into contact with and be driven by the friction disks, or, in other words, as the operator moves the turret on its centre the friction roll of the drill spindle last used gradually passes out of contact with the friction disks, so that only the spindle that is actually in position for use revolves at one time, the others all remaining stationary.

The feed to the drill is given by operating the lever L, which operates a pinion engaging in the rack seen on the table spindle. At N are nuts acting as stops to govern the amount of table motion to suit the work, but in addition to this each drill spindle is provided with a stop (one of which is seen at S) to govern the depth of hole drilled by that particular spindle. The machine is used for tapping as well as for drilling, countersinking, etc.

Fig. 381 represents Slate's sensitive drilling machine, in which the lower bearing for the live spindle is carried

Fig. 380.

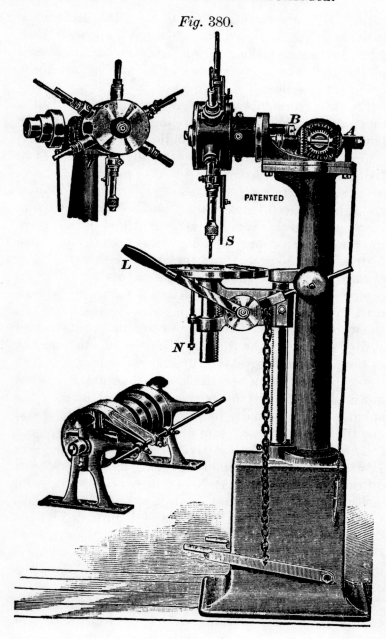

MACHINE TOOLS. 461

in a head H that fits to a slide on the vertical face of the frame, so that it may be adjusted for height from

Fig. 381.

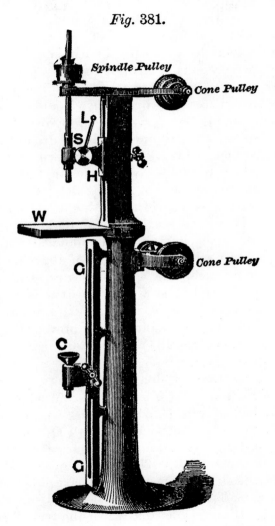

the work table W to suit the height of the work. L is a lever operating a pinion engaging a rack on the sleeve S to feed the spindle. The table W swings out of the

39*

way, and a conically recessed cup chuck C is carried in a bracket fitting into a guideway in the vertical bed G. The cone of the cup chuck is central to or axially in line with the live spindle, hence cylindrical work may have its end rested in the cone of the cup chuck, and thus be held axially true with the live spindle.

Fig. 382 represents a drilling machine by Prentice Brothers, of Worcester, Massachusetts. Motion for the cone pulley A is received by pulleys B and is conveyed by belt to cone pulley C, which is provided with back gear, as shown; the driving spindle D drives the bevel pinion E, which gears with the bevel-wheel F, which drives the drill spindle G by means of a feather fitting in a keyway or spline that runs along that spindle. Journal bearing is provided to the upper end of the spindle at H and to the lower end by bearings in the head J, which may be adjusted to stand at and be secured upon any part of the length of the slideway K. By this arrangement the spindle is guided as near as possible to the end L, to which the drill is fixed and upon which the strain of the drilling primarily falls. This tends to steady the spindle and prevent the undue wear that occurs when the drill spindle feeds below or through the lower bearing.

The feed motions are obtained as follows:

On the drill spindle is a feed cone M which is connected by belt to cone N, which drives a pinion O, that engages a gear P upon the feed spindle Q, which has at its lower end a bevel pinion, which drives a bevel gear upon the worm-shaft R. The worm shown on R drives the worm-wheel S, whose spindle has a pinion in gear with the rack T, which is on a sleeve U on the drill spindle G. It is obvious that when the rack T is operated by its pinion the sleeve U is moved endways, carrying the feed spindle with it and therefore feeding the drill to its cut, and that as the feed cone M has

Fig. 382.

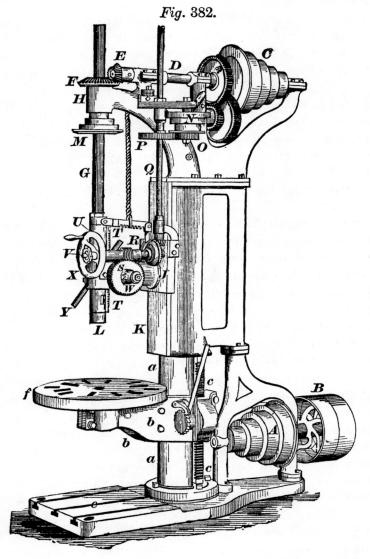

three steps there are three different rates of automatic feed.

To throw the self-feed into or out of action the following construction is employed:

The worm-wheel S has on its hub face teeth after the manner of a clutch, and when these teeth are disengaged from the clutch sleeve W the worm-wheel S rides or revolves idly upon its shaft or spindle, which therefore remains at rest. Now the clutch sleeve S has a feather fitting to its spindle or shaft, so that the two must, if motion takes place, revolve together; hence when W is pushed in so as to engage with S, then S drives W and the latter drives the spindle, whose pinion operates the rack T.

A powerful hand feed to the drill spindle is provided as follows:

The worm-shaft R is hollow, and through it passes a rod having at one end the hand nut V and at the other a friction disk fitting to the bevel gear shown at the right-hand end of the worm-shaft. This friction disk is fast upon the worm-shaft and serves to lock the bevel gear to the worm-shaft when the nut V is screwed up, or to release it from that shaft when V is unscrewed.

Suppose, then, that V is unscrewed and shaft R will be unlocked from the bevel wheel and may be operated by the hand wheel X, which is fast upon the worm-shaft, and, therefore, operates it and worm-wheel S, so that W being in gear with S the hand feed occurs when X is operated and V is released. But as the motion of S is, when operated by its worm, a very slow one, a second and quick hand feed or motion is given to the spindle G as follows, this being termed the quick return, as it is mainly useful in quickly removing the drill from a deep hole or bore:

The spindle carrying S and W projects through on the other side of the head J and has at its end the lever Y; hence W being released from S, lever Y may be operated, thus operating the pinion that moves rack T, one revolution of Y giving one revolution to the pinion, both being on the same shaft or spindle.

The work is carried and adjusted in position beneath the drill as follows:

The base of the column or frame is turned cylindrically true at a, and to it is fitted a knee b, which carries a rack c. The knee b affords journal bearing to a spindle which has a pinion gearing with the rack c, and at the end of this spindle is a ratchet-wheel d, operated by the lever shown. A catch may be engaged with or disengaged from ratchet d. When it is disengaged the lever may be operated, causing the pinion to operate on rack c and the knee b to raise or lower on a according to the direction in which the lever is operated. As the knee b carries the rack the knee may be swung entirely from beneath the drill spindle and the work be set upon the base plate e if necessary, or it may be set upon the work table f, which has journal-bearing in the knee b, so that it may be revolved to bring the work in position beneath the drill.

What is termed *a radial drill, or radial drilling machine*, is one that has an arm capable of swinging about upon the column that supports it, this arm carrying the head and feed motions for the drill. Fig. 383 represents a radial drilling machine, the arm of which is pivoted at a, so that it can swing in a horizontal plane; e is a slideway for the saddle f carrying the drill spindle and the feed motion, the hand feed being through the medium of the hand-wheel g.

Radial drills are especially useful for heavy work, as it is easier in that case to move the radial arm to position over the work than it is to move the work under the drill.

Keyway cutting machines, or keyway cutters, as they are generally termed, are much more expeditious in cutting keyways in the bores of wheels, pulleys, etc., than the

466   COMPLETE PRACTICAL MACHINIST.

slotting machine, and furthermore, when properly constructed, produce better work.

Fig. 384 represents the Morton Manufacturing Com-

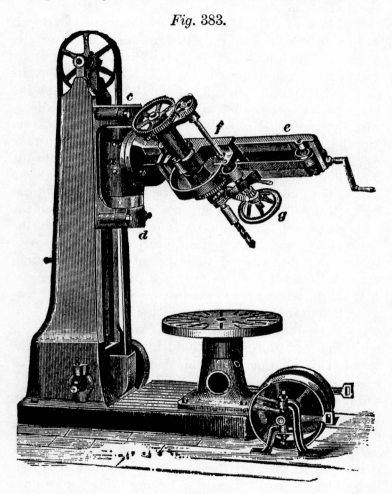

Fig. 383.

pany's keyway cutter, which cuts on the downward stroke and is a very efficient tool. The cutter bar C may be set to operate vertically or to operate at the angle necessary to give any required degree of taper to

MACHINE TOOLS. 467

the keyway, and can, at the same time, be operated to keep the tool clear of the cut on the upward, which is the non-cutting, stroke.

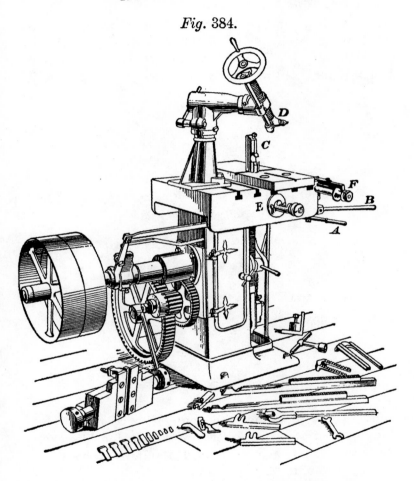

Fig. 384.

The handle A is for starting and stopping the machine. Handle B moves the cutter bar C up to its cut. E is a micrometer for regulating the depth of the keyway to be cut, and F is for regulating the feed, which is done as follows:

During the upward or return stroke of the cutter handle B is operated to move the cutter bar, and therefore the cutting tool away from the cut, and the stop nut at F is moved sufficiently to give the amount of tool feed necessary for the nut cut. As soon as the downward stroke commences handle B is pulled forward as far as the stop nut at F will permit it to go. D is merely a clamp for holding the work down upon the work

Fig. 385.

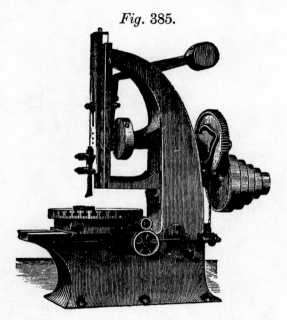

table. These machines are made in several sizes, including a portable one for use on wheels, too large and heavy to be chucked or moved about, or upon fixed housings or columns.

Slotting machine. In the slotting machine the tool is carried at the lower extremity of a ram or bar that operates vertically, the work being held upon a table lying in an horizontal plane beneath the ram.

This table is capable of being revolved upon its axis

or traversed in two directions, one at a right angle to the other.

Fig. 385 represents a slotting machine by William Sellers & Company, of Philadelphia, the construction

Fig. 386.

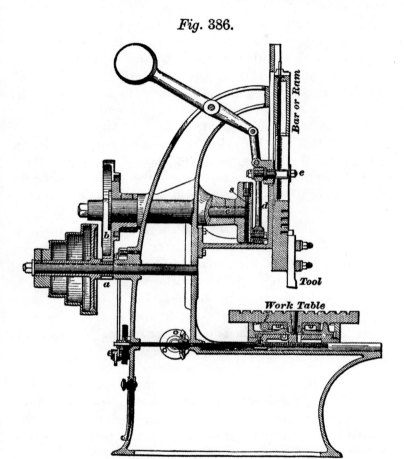

being more clearly seen in Fig. 386. The cone pulley drives a pinion *a* driving a spur gear *b* on the shaft *c* operating a Whitworth quick return motion from which the rod *d* drives the ram, which may be adjusted for height from the work table by means of adjusting the position of the driving pin *e*.

The longitudinal feed of the work table is operated by the screw shown beneath the work table in the sectional view, Fig. 386; the transverse or cross-feed is automatic, as is also the circular feed of the table.

The slotting bar is counterbalanced; this effectually prevents jar in running, the lost motion being all taken up by the counterweight in the direction of the force exerted in making the cut. The compound table is provided with a circular table, operated by wheel and tangent screw with self-operating feed. An important feature of this machine is the arrangement of its feed motion, which insures the feed always occurring at the top of the stroke and never during the cut. The great advantage of this will be manifest when it is remembered that should the feed occur at the lower end of stroke the rigid tool will drag back with a pressure due to the amount of feed. The working handles to operate feed by hand are on all these machines, even on the largest size, within easy reach of the workman.

Bolt cutting machines or bolt cutters are employed to cut the threads upon bolts. Such machines are made single and double, that is to say, with a single head or a double head. Each head contains the dies or chasers, which are provided with means to close them to cut the thread to the required diameter, and in the cases of simple machines to run backward to withdraw the dies from the bolt, while in improved machines the dies are opened automatically so that the bolt can be withdrawn as soon as its thread is completed. The bolts are held in jaws or chucks that are moved by hand-wheels operating right- and left-hand screws so that the jaws open and close equally, and the bolts will be held in line with the axis of revolution of the thread-cutting dies. The bolts are moved up to the dies sometimes by **levers and at others** by a rack and pinion motion.

MACHINE TOOLS. 471

Fig. 387.

Fig. 387 represents a National bolt cutter in which

the jaws for holding the bolt heads are operated by the left hand hand-wheel and the forward motion is operated by the hand-wheels on the right, the rack and pinion motion being clearly seen. The automatic motion for opening and closing the dies is shown in Fig. 388, the construction being as follows:

Above the forward main spindle bearing is fixed a bracket A, to which is pivoted the ring lever designated by the letter H. This ring carries blocks or fingers at the point (2) which engage with a sliding ring on the head,

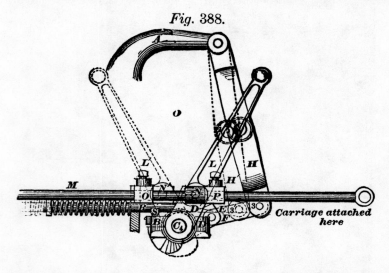

Fig. 388.

and is moved back and forth at the bottom by means of the link F, which connects with a short lever arm which is on the shaft C. The rod R having the spiral spring at one end is also connected with the ring H, and tends constantly to draw it to the left and open the head. When the parts are in the position shown by the full lines, the three centres joining the lower end of the ring H, the link F and the short lever arm are in line, and the parts are thereby locked in position. The lever L is attached to the shaft C_4, and carries upon its side a block D', through which passes the rod M, carrying two tappet collars O and P. This rod is attached to the carriage, and, as it moves up, the collar P comes in contact with the block D, and moves the lever L to the left. As soon as sufficient movement takes place to raise the end of the link F a short distance, the coiled spring draws the lever rapidly backwards and opens the head. The set screw S, which has been replaced with a slotted stop, is adjusted to prevent the lever L moving too far to the right, or beyond the point at which the locking takes place. When the carriage is drawn back, the collar O acts to close the head. By moving the collars out of the way, the head can be opened and closed by hand.

40*

CHAPTER XX.

TO CALCULATE THE SPEED OF WHEELS, PULLEYS, ETC.

MULTIPLY the speed of the driving-wheel by the number of teeth it contains, and divide by the speed required by the driven wheel.

Example 1.—If a wheel contains 50 teeth and makes 25 revolutions per minute, what number of teeth must a wheel contain to gear into it and make 125 revolutions per minute?

$50 \times 25 = 1250 \div 125 = 10.$ *Ans.* 10 teeth.

Example 2.—A wheel contains 90 teeth and makes 120 revolutions per minute, how many teeth must a wheel contain to gear into it and make 240 revolutions per minute?

$90 \times 128 = 10800 \div 240 = 45.$ *Ans.* 45 teeth.

Example 3.—A wheel contains 45 teeth and makes 240 revolutions per minute, how many teeth must a wheel geared into it contain to run 120 revolutions per minute?

$45 \times 240 = 10800 \div 128 = 90.$ *Ans.* 90 teeth.

In the case of pulleys or band-wheels the rule is the same, except that the diameter of the wheel is taken instead of the number of teeth.

Example 1.—A driving-wheel makes 120 revolutions per minute and is 24 inches in diameter, what size pulley must I employ to obtain 60 revolutions per minute?

$120 \times 24 = 2880 \div 60 = 48.$ *Ans.* 48 inches.

Example 2.—A driving-wheel four feet in diameter makes 245 revolutions per minute, what size pulley must I use to obtain 45 revolutions?

SPEED OF WHEELS, PULLEYS, ETC. 475

Diameter of wheel in inches.

$48 \times 245 = 11760 \div 45 = 248.$ *Ans.* 248 inches.

Example 3.—A driving-wheel makes 182 revolutions per minute, and is 15 inches in diameter, what size pulley do I require to make 145 revolutions?

$182 \times 15 = 2730 \div 145 = 18.965.$

Ans. A pulley $18\frac{96}{100}$ in. in diameter.

Another rule which will answer, whether we employ a single pair or two pair of pulleys, is as follows:

Divide the speed you require to run by the speed of the driving-shaft, and the quotient will be the proportion between the revolutions of the driving-shaft and the revolutions required. Then take any two numbers that will, when multiplied together, form a sum equal to that proportion, and one of such numbers will form the relative sizes for one of the pairs of pulleys, and the other of such numbers will form the relative sizes for the other pair of pulleys.

Example.—It is required to run a machine 1200 revolutions per minute, the driving-shaft makes 120 revolutions per minute, what sizes of pulleys shall be used?

Revolutions required.	Revolutions of driving-shaft.	Proportion of speed.
1200 ÷	120	= 10
Then 5 ×	2	= 10
or 4 ×	$2\frac{1}{2}$	= 10
or $3\frac{1}{3}$ ×	3	= 10

So that the proportion being ten to one, we may use two wheels of any sizes, providing that the one on the driving-shaft is ten times as large as the one on the machine: or since $5 \times 2 = 10$, we may place on the driving-shaft a pulley, say five feet in diameter, and belt it to one a foot in diameter, forming the proportion between the first pair of pulleys of five to one. Our next pair of pulleys must be two to one, that is to say, we may use a two foot and a one foot pulley, a four and a two foot, or any others, so that one is twice as large as the other. Or again, since $4 \times 2\frac{1}{2} = 10$, we may use a four foot pulley, or the driving-shaft belted to a foot pulley, or a 24 inch one belted to a 6 inch one, the proportion in either case being 4 to 1, then for the second pair we may employ a 12 and a 6 inch, or a 24 and a 12, or any two pulleys we may have on hand, so that one is twice as large as the other. It is obvious that when the speed required is greater than that of the driving-shaft, the large pulleys are the driving and the small ones are the driven pulleys.

CHAPTER XXI.

HOW TO SET A SLIDE VALVE.

In setting a slide valve, we are confronted with the following considerations:

Our object is to cause the admission to expansion of and exhaust from the cylinder of the steam equal for each stroke. This, however, we are unable to attain, because of the angle of the connecting rod. If we set the valve so that the exhaust commences in the same relative position of the piston at one stroke as compared to the other, the valve will not admit the steam to the cylinder in the same relative position of the piston at one end of the stroke as compared to the other—that is to say, the valve being set so that the exhaust will take place when the piston has moved an equal number of inches of the stroke at either end of the cylinder, the steam port, when the crank is on the respective dead centres, will be wider open at one end than at the other. Then, again, if we set the valve so that the exhaust commences at an equal part of the stroke at either end of the cylinder, the exhaust port will be wider open when the crank is on one dead centre than it will when the crank is on the other dead centre. Whereas, if we set the valve so that the steam port at each end is open to an equal amount when the crank is on either respective dead centre, the exhaust port will also be open at each end to an equal amount when the crank is on the dead centre. Thus, by setting the valve so that it has an equal amount of lead at each end of the piston stroke, the exhaust ports will be open to an equal amount when the piston commences its return stroke at either end of the cylinder. It is always, therefore, preferred to set the valve so that it shall have lead to an equal amount when the crank is on the dead centre.

If the stroke of a valve is made to be twice the width of the steam port added to twice the amount of the lap, it will be found that the steam port at the end of the cylinder farthest from the crank will not open fully, while that nearest to the crank will have the valve travel past it to an equal amount; the degree of this difference becoming greater as the amount of the lap is increased, and hence, as the amount of the lead or angular advance of the eccentric becomes greater.

The exhaust of the steam will, when the lead of the valve is equal at each end of the stroke, take place earlier in the stroke in the front end of the cylinder—that is, the end the farthest from the crank—than it will at the back end, while the steam will work expansively during a greater part of the stroke when it is in the back end than will be the case when it is in the front end of the cylinder. Having noted these facts, we may proceed to set a valve, the first operation being to carefully remove, by blowing or washing out, any filings or scrapings that may have lodged in the ports while operating upon. The valve-seat, or the cylinder, the bore of the cylinder, the valve-face, and the face of its seat, having been carefully cleaned and then oiled, we may connect the various parts, and find and mark the dead centres or dead points of the stroke as follows:

In Fig. 389 A represents the guide-bar, B the guide-block, C the fly-wheel, D the crank, E the eccentric, and F the centre-line of the connecting-rod of an engine intended to run in the direction of the arrow.

Giving the wheel a turn or two in the direction in which it is intended to run, we allow it to come to rest, so that the motion-block B will be at very nearly the end of its

HOW TO SET A SLIDE VALVE. 479

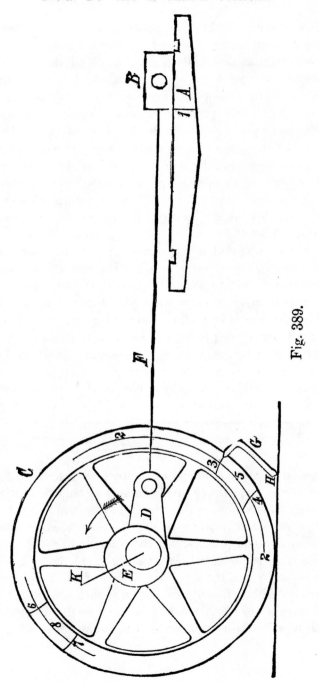

Fig. 389.

stroke on the guide-bar A, and then placing the edge of a straight edge along the end of the guide-block B, the straight edge at the same time overlapping the face of the guide-bar, we mark on the face of the latter the line 1, which will thus be quite even with the end-face of the guide-block. We then (after chalking it to make the marks show plainly) mark on the face of the wheel the line 2, which should be true with the centre of the main shaft, but which can be marked from the rim of the wheel with a pair of compass calipers, providing that rim has been trued up in the lathe. We next, with a piece of iron wire or rod bent as shown by G, mark at some fixed point, such as shown at H; make a centre-punch mark, and resting one end of the scriber G in the fixed centre-punch mark, we scribe with the other end upon the edge of the wheel the line 3, as shown in the illustration. Our next operation is to move the wheel forward in the direction in which it is to run, so that the crank will move to the dead centre, and the guide-block will leave the line 1, as shown in Fig. 390, and the motion of the wheel being

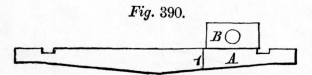

Fig. 390.

continued, the guide-block will return to the mark 1, the wheel being moved very slowly indeed, so that there will be no trouble so to move it that the end of the guide-block will come to rest exactly fair with line 1. If, by chance, the end of the guide should move past the line, the wheel should be turned well back—that is to say, back to the end of the stroke—and again moved slowly forward till it comes fair with the line 1. The object of this is that the guide-block shall always approach the line moving in the direction in which it will, while performing that stroke, move when the engine is at work, so that all the working parts will be brought to a bearing in the direction in which they will bear when at work, and hence any spring or lost motion in any of the parts will not affect the setting of the valve.

Suppose, for instance, there was even a trifling amount of play in the eccentric or any of the bolts, and that the end of the guide-block on its return stroke having moved a trifle past the line 1, we move the wheel backward a trifle to correct the error, thus making the block approach the line from the opposite direction to what it will approach it when at work and travelling on that stroke, then part of the movement of the wheel will have been lost (so far as the movement of the guide-block is concerned), having been expended in taking up the lost motion. It makes no difference if the engine is to run both ways, for in that case we observe the same precaution in moving the wheel, setting the valve in the forward gear, and then trying it in the backward gear, and dividing the difference, if there be any.

To proceed, then, the guide-block having returned even with the line 1, we take our wire scriber, rest one end in the fixed point, and with the other end mark on the edge-face of the wheel line 4, which will then occupy the place that line 3 does in our engraving. Our next duty is to find the centre between lines 3 and 4 as shown in Fig. 391, which

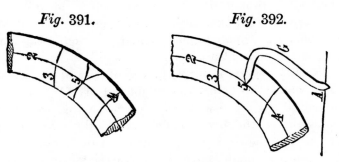

Fig. 391.   *Fig.* 392.

we obtain from lines 3 and 4, and which we mark with a

fine centre-punch mark, as shown at 5. And it will readily be perceived that if we move the wheel round so that the scriber G resting in the fixed centre-point as shown in Fig. 392, the end will be true with centre-punch mark 5, the motion-block, and hence the piston and crank, will be exactly on the dead centre at that end of the stroke.

We next move the wheel around, so that the guide-block will be nearly at the end of its stroke at the opposite end of the guide-bar, and mark a line occupying the same relative position at that end as line 1 does at the other end, and repeat the whole previous operation, thus marking new lines corresponding to the lines 2, 3, 4, and 5, but on the opposite diameter of the wheel; thus we shall obtain the lines 6 and 7 in Fig. 389, and from them the centre-punch mark 8, which will serve the same purpose at that end of the stroke as does the centre-punch 5 at the opposite end.

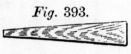

Fig. 393.

Our next procedure is to provide a small wooden wedge, such as is shown in Fig. 393, making one end of a thickness equal to about half the amount of lead it is intended to give the valve, and the other about twice as thick as the intended amount of lead, its length being about three inches. We then move the wheel in the direction in which it is intended to run, until the scriber, one point resting in the fixed centre-punch mark, the other will be exactly even with centre-punch mark 5, and the engine will be on the forward or front dead centre.

We are now ready to set the eccentric, and the question arises, which way the engine ought to run? We have performed all our operations thus far with a view to have the engine run in the direction denoted by the arrow in 389; for had the engine been intended to run in the opposite direction, we should have drawn line 1 while the guide-block was in the same position on the guide-bar, but with the

HOW TO SET A SLIDE VALVE. 483

crank on the other side of the same dead centre. The direction of arrow C, *Fig.* 394, is the correct one in which an engine should run, if circumstances permit, for the following reasons: in Fig. 394, if we suppose A to represent the centre-line of the connecting-rod, and the engine to be running in the direction of the arrow B, then the strain on the connecting-rod will be in a direction tending to compress it, the strain on the guide-bar being in the direction to force the cross-head guide-blocks down upon the guide-bars, and hence to produce the most friction; whereas if the engine was running in the direction denoted by the arrow C, the strain upon the connecting-rod will be one in a direction to pull it apart, and hence to lift it and the cross-head guide-blocks from the guide-bars, so that the centre line of the connecting-rod would stand in the direction of the line D. When the crank is on the other side of the dead centres, the same effect in either case is produced. Now it is quite true that so long as the guide-blocks fit to the guide-bars, the rod cannot move in any

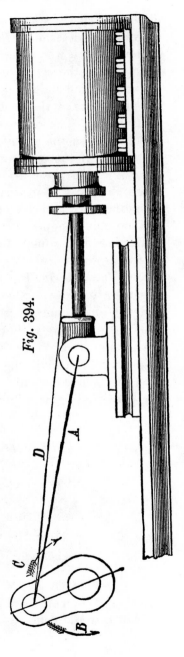

Fig. 394.

direction, but the spring of the various parts, the direction of which is determined by the direction of the strain, is sufficient, even when the engine is new, and hence, there is no play in the guide-blocks to (if the engine is running in the direction shown by the arrow C in Fig. 394) relieve the guide-bars of the friction due to the weight of the connecting-rod. The only objection to be advanced against running an engine in the direction to relieve the slides of the weight of the connecting-rod is, that in such case the wear of gibs of the cross-head will be mainly on the underneath side, and that, therefore, the play should be taken up on that side, and the set screws provided for that purpose will, on many engines, be somewhat difficult to get at.

Having determined, then, the direction in which the engine is to be run, we place our eccentric so that its throw line (K, in Fig. 389) will stand sufficiently in advance (in the direction in which the engine is to run) to let the front port open to the required amount of lead, and fasten it there with the set-screw. We then measure the amount of lead there is on the valve when the eccentric is so set, by chalking the faces of the wooden wedge, and inserting it in the opening or lead of the port, as shown in Fig. 395, putting it in between the edge of the valve and

Fig. 395.

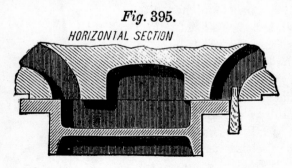

HORIZONTAL SECTION

the edge of the port until it is a snug fit, but not forcing it in (which would compress the wood). When the wedge is home it should be moved edgeways, and then taken out, and the steam port edge will have left a mark on the wedge, evidencing how far the wedge entered, and therefore the precise amount of the lead at that end. We then move the wheel forward until the crank is on the other dead centre—that is to say, until the centre-punch mark No. 8 comes exactly even with the scriber point G—and try the wedge in the back port, and if it enters at the same distance as it did at the front port, the valve is set.

If, however, there is found to be more lead at one end than at the other, it demonstrates that the eccentric rod is not the correct length; if the front port has the most lead, the eccentric rod is too short, and *vice versâ*. Locomotives and other engines, in which the height of the main shaft will vary in its relation to the height of the cylinder according to the weight of the engine or its load (as of fuel, water, etc.), should have their valves set with the load on and the engine in its working position, or moved along the rail, instead of revolving the wheel with the engine lifted off the rail.

It is an excellent plan after an engine valve is set, to take the scriber G, in Fig. 389, and making it exactly 6 inches long from point to point (so that its length may always be known and remembered) rest one point on some part of the steam-chest, and in a centre-punch mark provided for the purpose, and with the other, mark a line on the slide-spindle, when the engine is on each dead centre. Then put a centre-punch mark on the slide-spindle. Thus with the gauge applied to the steam-chest and slide-spindle, the valve may be set, in cases of necessity, without taking the steam-chest cover off. It is better, however, to remove the steam-chest cover if circumstances permit.

CHAPTER XXII.

PUMPS.

PUMPS are commonly divided into three classes, the suction pump, the force pump, and the suction and force pump.

SUCTION PUMPS.

A suction pump causes water to raise itself, by relieving its surface of the pressure of the column of air resting upon it. The principle upon which it acts may be explained as follows:

The surface of all water exposed to the air has the pressure of the air or atmosphere resting upon it; if, therefore, one end of a pipe or tube be lowered into water, and the other end be closed by means of a valve or other device, and the air contained in the pipe be drawn out, it is evident that the surface of the water within the pipe will be relieved of the pressure of the atmosphere; and there will be no resistance offered to the water to prevent its ascending the pipe. The water outside of the pipe, still having the pressure of the atmosphere upon its surface, therefore forces water up into the pipe, supplying the place of the excluded air. The water inside the pipe will rise above the level of that outside of the same in exact proportion to the amount to which it is relieved of the pressure of the air, so that, if the first stroke of a pump reduce the pressure of the air contained in the pipe from 15 lbs. on the square inch (which is its normal pressure) to 14 lbs. per inch, the water will be forced up the pipe to the distance of about $2\frac{1}{4}$ feet, because a column of water

PUMPS.

an inch square and 2¼ feet high is equal to about 1 lb. in weight.

It is evident that, upon the reduction of the pressure of the air contained in the pipe from 15 to 14 lbs. per square inch, there would be (unless the water ascended the pipe) an unequal pressure upon its surface inside as compared to that outside of the pipe; but in consequence of the water rising 2¼ feet in the pipe, the pressure on the surface of the water, both inside and outside, is evenly balanced (taking the level of the outside water to be the natural level of the water inside), for the pressure upon the water exposed to the full atmosphere will be 15 lbs. upon each square inch of its surface; while that upon the same plane, but within the pipe, will sustain a column of water 2¼ feet high (weighing 1 lb.) and 14 lbs. pressure of air, making a total of 15 lbs., which is, therefore, an equilibrium of pressure over the whole surface of the water at its natural level.

If, in consequence of a second stroke of the pump, the air pressure in the pipe is reduced to 13 lbs. per inch, the water will rise up it another 2¼ feet, and so on until such time as the rise of the column of water within the pipe is sufficient to be equal in weight to the pressure of the air upon the surface of the water without; hence we have only to determine the height of a column of water necessary to weigh 15 lbs. per square inch of area at the base of the column to ascertain how far a suction pump will cause water to rise, and this is found by calculation or measurement to be a column nearly 34 feet high. It becomes apparent, then, that, however high the pipe may reach above the water level, the water cannot rise more than 34 feet up the pipe, even though all the air be excluded within the pipe, because the propelling force, that is, the atmospheric pressure, can only raise a column of water equal in weight to itself. It is found, however, in practice, to be an excellent suction pump which will raise water thirty feet.

From this it will be perceived that the terms "drawing water" and "suction pump" do not accurately represent the principles upon which this pump performs its duty; and it would be much more proper to call it a "displacement pump," since its action is simply to enable the water to rise by displacing the air from its surface.

The duty of this pump is, therefore, in the first place, to extract the air from the suction pipe, and, in the second place, to discharge the water from its barrel through the medium of valves in such a manner that the column of water in the suction pipe is at all times entirely excluded from the pressure of the atmosphere.

FORCE PUMPS.

A force pump is one by means of which the water is expelled from the pump barrel and through the delivery pipe by means of the mechanical force applied to the pump piston or plunger; the amount of power required to drive such a pump will, therefore, depend at all times upon the height to which the water is required to be forced. When a pump is arranged to draw the water, and force it after it has left the pump barrel, it is termed a suction and force pump; but if the water merely flows into it in consequence of the level of the water supply being equal to or above that of the top of the pump barrel, it is termed simply a force pump. Hence a suction pump performs its duty in causing the water to rise to the pump, a force pump is one which performs its duty in expelling water from its barrel, and a suction and force pump is one which performs both duties alternately.

All pumps require a suction and a discharge valve, the suction valve being so arranged as to open to admit the water into the pump barrel while the pump piston or plunger is receding from that valve, and to close as soon as the plunger stops or reverses its motion. The delivery valve is so arranged that it closes as the pump plunger or piston recedes from it, and opens when the same approaches it. When, therefore, the pump piston recedes from the suction valve, the latter opens and admits the water; and when the piston reverses its motion, the suction valve closes, and the descent of the pump piston forces the water through the delivery valve, that being its only possible mode of egress from the barrel of the pump.

The arrangement of the valves may be the same for a force as for a suction pump (although it is advisable, in some cases, to place an additional valve to a force pump to prevent the pump piston from receiving the force of the water in the delivery pipe), the only difference being that the water is permitted to flow freely away from a suction pump, whereas it is confined to the delivery chamber or pipe in a force pump, so as to force it to the required height or pressure, as the case may be.

PISTON PUMPS.

A piston pump is one in which the water is drawn or forced by means of the piston fitting the barrel of the pump air-tight, which is most commonly done by providing the piston with two cupped leathers, formed by being pressed in a die made for the purpose. The leather is soaked in the water before being placed in the die, and is allowed to remain in the die until it is dry, when it will be sufficiently hard to admit of being turned in the lathe.

The capacity of a piston pump is its area multiplied by the length of its stroke; but it must be remembered that all pumps throw less water than their capacity, the deficiency ranging from 20 to 40 per cent., according to the quality of the pump. This loss arises from the lift and fall of the valves, from inaccuracy of fit or leakage, and in many cases from there being too much space between the valves and piston or plunger.

A plunger pump is one in which a plunger is used in place of a piston, the gland through which the plunger moves serving as its guide, and also keeping it air and water-tight. The plunger is made smaller in diameter than the bore of the pump barrel, so that the capacity of such a pump is the area of the end face of its plunger multiplied by the length of its stroke, because the pump acts by reason of the displacement caused by the plunger entering the barrel. Pump-plungers should always be draw-filed lengthways to prevent them from wearing away the packing so rapidly. It is always advisable in this kind of pump to allow as small an amount of space between the plunger and barrel as possible, for the following reason: When the plunger becomes worn, it is necessary to turn it up again in the lathe, thus reducing its diameter. The result is that there is so much air in the pump, between its barrel and the plunger, that it expands as the plunger leaves the barrel and is merely compressed by the plunger returning, so that the pump becomes very ineffective, and finally ceases to pump at all. If the pump, in such a case, be primed with water each time it is started, it may draw water, but not to its full capacity, as the air will remain in the pump barrel until such time as it may become absorbed by the water.

Suction valves for all pumps should be made as large in area as it is possible to get in, so that they will not require to lift much to admit the water to the pump: since it is evident that, when the piston or plunger commences to descend and the suction valve to close, the water passes back through the suction valve until it is closed, thus diminishing the effectiveness of the pump, and, further, causing the valve to close with a blow which proves very destructive to the valves, especially of fast-running pumps.

The area of the opening of a suction valve must be at least equal to the area of the suction pipe, whose area is determined by the following principles: Water will not flow through a suction pipe in a solid stream at a greater speed than that of 500 feet per minute. It follows, then, that, the quantity of water the pump is required to throw being determined, the suction pipe must be of such a size that 500 feet of it will hold such quantity.

If the suction pipe be any smaller than that size, the pump will not be fully supplied with water; and the piston or plunger travelling faster than the supply of water follows it, there is, when it arrives at the end of its suction stroke, a partial vacuum in the pump barrel, which keeps the suction valve open. When the piston or plunger has descended until it strikes the water again (the suction valve not having yet closed), the water, descending with the piston, strikes the suction valve with a blow, which, as before stated, gives a backward impetus to the water in the suction pipe, and closes the valve with a blow very destructive to it; especially is this the case in a force pump or a fast running steam pump, in which latter case the steam piston accelerates in speed (when the pump piston has the partial vacuum referred to in it) because not only is the steam piston relieved from performing any duty, but it is assisted by the vacuum; so that it accelerates its speed greatly until the piston strikes the water in the pump barrel, which it will do with such force as to very probably break some weak part of the engine or pump, or cause the crossheads or piston to become loose. If the working parts of any pump are accurately fitted, it will deliver more nearly its full capacity of water when running slowly.

An air-chamber placed in the suction side of any pump causes a better supply of water to the pump by holding a body of water near to it, and by making the supply of water up the suction pipe more uniform and continuous. Air-chambers should be made as long in the neck as possible or convenient, so that the water, in passing from the pump barrel to the delivery pipe, shall not be forced up into the chamber at each stroke of the pump; for the air in the chamber becomes gradually absorbed by the water. If fresh water is continually passing into and out of the chamber, the air in it will soon become absorbed, and water will supply its place; but if the air-chamber has a long neck, the water at its highest level in the chamber will remain there unchanged by the action of the pump, and will become impregnated with air, thus diminishing its propensity to absorb any more; and although the air will finally become all absorbed out of the air-chamber, the process is a very much slower one, the air-chamber being so much the more effective, and its elasticity, imparting a steady flow of water from the delivery pipe, being unimpaired.

Pumps whose pistons revolve are subject to the same defects from inequality of wear as are rotary engines, but the results are not so keenly experienced, because water will not leak through so rapidly nor to so serious an extent as steam, and, further, because the leakiness of the pump may be compensated for by an increase of the rotative speed of the piston.

Water will not, however, as before stated, flow through the suction pipe at a greater velocity than 500 feet per minute; so that, if the pump performs more revolutions than are requisite (according to its capacity) to carry off more than the quantity of water contained in 500 feet of its suction pipe, the power used in running those extra revolutions is lost, inasmuch as they are superfluous except for the purpose of compensating for the defects in the construction or leakiness of the pump, in which case the excess of speed becomes a necessary evil.

In actual practice it is found that the efficiency of a pump is appreciably increased by increasing the size of suction pipe to a diameter sufficiently large, that the water requires to travel through it at a speed of not more than 200 feet per minute. It is also found that the effective-

PUMPS.

ness of a pump varies with the height it draws the water, or in other words, the length of the suction pipe.

Prominent among the causes of the loss of efficiency in pumps appears to be the want of a comparatively large body of water close to the suction side of the pump. Supposing a single acting pump, such as plunger pumps usually are, to have a suction pipe capable of supplying as much water flowing through it, at a speed of 500 feet per minute, as the pump delivers per minute. In the first place, the water only flows through the suction pipe at and during the up stroke, so that the water will, in such a pump, have to pass through the pipe at a speed equal to 1000 feet per minute. So that the suction pipe for a single acting pump should be of such a size that it will deliver as much water flowing through it at a speed of 250 feet per minute, as the pump delivers per minute, in which case the water in the suction pipe will, while actually in motion, move at a speed of 500 feet per minute.

If, however, the pump is a double acting one, the suction pipe may be of such size that it will deliver, flowing through it at a speed of 500 feet per minute, as much water as the pump will deliver per minute, or in other words, the suction pipe should bear the same proportion in size to a single acting as to a double acting pump.

If a pump makes 60 strokes per minute, and the water flows through the suction pipe at the rate of 500 feet per minute, it is obvious that each stroke of the pump will draw the water from a length of over 8 feet of the pipe. Now it is a good pump which will deliver 80 per cent. of its capacity; allowing a loss of 10 per cent. for the lift and fall of the valves, we still have a loss of 10 per cent. unaccounted for, and it is self-evident that the column of water entering the pump must be broken somewhere, to a partial extent, at least, and it is reasonable that the break occurs close to the pump, because the water close to the pump has not the friction on the sides of the pipe tending to hold it back. Now, if a pump was provided with a reservoir (similar to a steam chest), and the suction valve was at the bottom of such reservoir, there would be very little liability of the pump failing to get a full supply of water, because the water in the reservoir would flow readily into the pump, and the break, if any, would be at the upper part of the reservoir, which break would, provided that the suction pipe entered the reservoir at the bottom, cause a continuous flow of water up that pipe, while at the same time the supply to the pump would not be appreciably affected by the break. An air-chamber on the suction side of a pump does not fulfil these conditions, because the orifice connecting the air-chamber to the suction valve is comparatively small, whereas the whole body of water in the proposed reservoir would be in full and unconfined communication with the suction orifice or port of the pump. If a large steam chest induces an initial pressure, to a steam cylinder, more approximate to that obtaining in the boiler, how much the more necessary is a similar chest or reservoir necessary to a pump, especially when it is considered how much greater is the inertia of water than that of steam.

INDEX.

ADJUSTABLE dies, 250, 251.
 Alloys, 237.
American chrome steel for tools, 221, 222.
Angle of top and bottom faces of tools, 39.
Angles of threading tools, gauge for testing, 90–92.
Annealing or softening, 236.
Arbors or mandrils, 120–122.
Asbestos for joints, 313.

BABBITT metal bearings, wear of, 235, 236.
 metal for fast-running journals, 237.
Bar and cutter, 168.
Barnes Co.'s 5 foot-power lathe, 414–416.
Bars, boring, 178–190.
Bay Chaleur grindstones, 347.
Bearings, Babbitt metal, for fast-running journals, 237.
 Babbitt metal, or of brass, wear of, 235, 236.
 cast-iron, wear of, 230.
Bell metal, 237.
Bits and reamers, 71.
 half-round, 166–168.
Blacksmith's tools, effect of blows in forging, 238.
Bolt cutter, National, 471–473.
 cutting machines or bolt cutters, 470–473.
Bolts upon targets, representing ship's armor, experiments with, 94.
Boring bar, conditions necessary in a, 178.
 bar cutters, 179.
 bar, for use in bores of a large diameter, 181.
 bar, importance of, 178.
 bar, keyway in, 180.
 bar, the position which the cutter should occupy toward the head of, 183–186.
 bars, 178–190.

Boring bars, small, 189, 190.
 light brass work, 79.
 links or levers, 144, 145.
 tool for brass, 78–81.
 tool holders, 82, 83.
 tools for lathe work, 71–83.
 tools for lathe work, shaping, 71.
 tools for wrought-iron, cast-iron, steel or copper, rake of, 79.
 tools, illustrations of the various forms of, 76, 77.
Boult's panelling and dovetailing machine, 210.
Boxes, fitting brasses to, 284, 285.
Brass, boring tool for, 78–81.
 cutting, speeds and feeds for, 70.
 for journal boxes, 237.
 for valves, 237.
 work, boring, 79.
 work, front tool for, 50–53.
 work, grooving tool for, 45.
 work, hand turning, 159, 160.
 work, roughing out tool for, 159, 160.
 work, scrapers for, 160–163.
 work, side tool for, 53, 54.
 yellow, 237.
Brasses, fitting to their boxes, 284, 285.
 letting together, to take up the wear, 323.
 lining up to set the key, 324.
 of a connecting rod, filing, 323.
Brown & Sharpe's milling machines, 453–458.
 plain milling machine, 453–455.
 tap, 248, 249.
 universal milling machine, 455–457.

CALCULATING the speed of wheels, pulleys, etc., 474–476.
Calipers, 90, 91, 264–267.

(495)

INDEX.

Carrier or dog for lathe work, a simple form of, 116, 117.
Case-hardening wrought iron, 229, 230.
Cast-iron, cutting speeds and feeds for, 70.
   in bearings or boxes, wear of, 230.
   piston rings, wear of, 230.
   slide valves, wear of, 230.
   under steam pressure, wear of, 232.
   use of square-nosed tools on, 38.
Cast-steel, to make very hard, 228.
Centre-drilling attachment for lathe work, 124, 125.
   by hand, 125, 126.
   combined drill and countersink for, 124, 125.
Centring lathe work, 122–127.
   machine, 124.
Change gear wheels for screws, to calculate, 94–102.
Charcoal for heating tools in hardening, 226.
Chaser, outside, for cast-iron or brass, 109, 110.
   for cutting wrought iron, 109–110.
   to make a, 107–115.
Chasers, proportions for, 108.
Chasing, hand, 104–107.
Chisels, 256–264.
Chrome steel for tools, 221, 222.
   steel, making very hard, 226.
Chuck a cross-head, to, 141–143.
   box body, 131.
   dog, 135.
   expanding, for holding piston rings, 140, 150.
   of the Russell Tool Co., 134.
   the, 412.
   the Horton two-jawed, 131.
   the Sweetland, 133, 134.
Chucking an eccentric, 136, 138, 139.
   hand-turned work, 153.
Chucks, lathe, 131–135.
   lathe, division of into two classes, 131.
   the large size of, 131, 132.
   universal or scroll, 135.
Clamp dog, 119.
Clamps, vise, 280, 281.
Clearance given by the bottom rake and the side rake of a tool, what dependent on, 40.

Clements' driver for lathe work, 116, 118.
Clutch, friction, of lathe, 424, 425.
Cocks, to rivet leaky plugs to, 304–308.
Color in tempering steel for tools, 222.
Compound or double screw-cutting gear, 97.
Cone pulley, to mark out a, 408–410.
Connecting rod, lining out, 396–408.
   rod of an engine, to fit a, 320.
   rod, to get the length of a, 319.
   rods, fitting, 314–324.
   rods, large, forging, 396, 397.
Copper, cutting speeds and feeds for, 70.
Counterbalancing work, 143, 144.
Countersink drill, a taper, 212, 213.
   drills, 212–214.
Crank of a horizontal engine, to ascertain when on its exact dead-centre, 319.
Cranks, turning, 140.
Crosshead, to chuck a, 141–143.
   to mark off a, 381.
   the bores of a, 141.
Cubical block, to mark a, 369–373.
Curve of an ellipse, to find the points through which it may be drawn, 364.
Cut, hard saw blades, to, 304.
Cutter and bar, 168.
   for cutting a thread, 55.
   stocks, 215.
Cutters, 214–218.
   and milling bar, 339–346.
   are steel bits, 214.
   bolt, 470–473.
   keyway, 465–468.
   of boring bars, 185, 186.
   recessing, 217, 218.
   tempering, 217, 218.
Cutting drifts, 326.
   edge in round-nosed tools, 36.
   edge of a tool, height of, relation of, to the work, 40–42.
   edge of a tool, pressure on, 72.
   edge of a tool, relation of the angle to which it stands to the cut, to the pressure, 73, 74.
   edge of a tool, the principles determining, 33–36.

Cutting edge of the tools of Whitworth's lathes, 41.
  edges of tools should be as near the tool post as possible, 41, 42.
  edges of twist drills, 197-200.
  off device, 59, 60.
  off machine, a form of lathe to cut rods or bars, 432.
  off machine of the Hurlburt Rogers Machine Co., 435.
  off machine of the Hurlburt Rogers Machine Co., valuable feature of, 435.
  off machine, slide-rest of, 434.
  off machine, spindle of, 432, 433.
  off machines, 432-435.
  off, parting, or grooving tools, 44-47.
  off tool holder, and steadying device, 59.
  out holes of a large diameter in sheet-iron, 216, 217.
  speed and feed, 64-70.
  speeds and feeds for brass, 70.
  speeds and feeds for cast-iron, 70.
  speeds and feeds for copper, 70.
  speeds and feeds for steel, 69.
  speeds and feeds for wrought iron, 69.
  speeds and feeds, tables of, 69, 70.
  surfaces of lathe tools, 84.
  tools, conditions affecting the shape of, 26.
  tools, effect of variation of shape or presentment of the work, 26.
  tools for lathes and planing machines, 25-63.
  tools for lathes, steel for, 25.
  tools for planing machines, 42.
  tools, rake of, 27-29.
  tools, steel for, 219.
  wedges, machine tools are, 29.
Cylinder posts and steam valves, marking out, 406-408.
Cylinders, fitting, 288-297.

DEAD-CENTRE of the crank of a horizontal engine, to ascertain its exact, 319.
Detrick & Harvey Machine Co.'s open side planer, 451, 452.
Diameter of work, and rake of a tool, relation of, 40.

"Diamond-pointed" tool, 28.
Dies, adjustable, 250, 251.
  and taps, 238-255.
  for use in hand stocks, 251-255.
Disc surfaces, revolving, experiment in regard to the wear of, 234.
Distance between the centres of two hubs of unequal height, 379-381.
Divide a straight line into a number of equidistant points, to, 367.
  a straight line into two equal parts, to, 367.
Dog or carrier for lathe work, a simple form of, 116, 117.
  the clamp, 119.
Dogs, chuck, 135.
Double eye, to line out a, 383-389.
  screw thread, to cut, 102-104.
Drift, using the, 327-329.
Drifts, 325-329.
  cutting, 326.
Drill and countersink, combined, for centre-drilling, 125.
  Farmer lathe, 204.
  flat, for enlarging and truing out holes, 168-171.
  holder, 170-171.
  tit, 202.
Drilling hard metals, 205, 206.
  in the lathe, 164-171.
  machine for drilling and tapping small holes, Quint's, 459.
  machine of Prentice Bros., 462-465.
  machine, radial, 465.
  machine, Slate's sensitive, 459-462.
  machines, 458-465.
Drills, common, used as countersinks, 213.
  countersink, 212-214.
  feeding, 202-205.
  flat, 201, 202.
  flat, defects in, 201, 202.
  ground by hand, testing for angle, 199, 200.
  machine-made shanks of, 204.
  pin, 211, 212.
  slotting or keyway, 206, 211.
  temper for, 204.
  twist, 196-201.
Driver for lathe work, Clements', 116, 118.

42 *

INDEX

ECCENTRIC, chucking an, 136–138.
    marking out an, 389–396.
Eccentrics, turning, 136.
Ellipse, to find the points through which the curve of, may be drawn, 364.
    to mark out an, 363, 364.
Emery cloth and paper, 128–130.
    wheel for grinding reamers, 172.
    wheel, use of, with milling cutters, 345, 346.
Engine guide bar, to mark out an, 373–378.
English grindstones, 347.
    or Whitworth standard for screw threads, 244.
    standard taps, flutes for, 246, 247.
Examination of work before marking out, 369.

FARMER lathe drill, 204.
    Feed and speed, cutting, 64–70.
Feeding drills, 202–205.
Feeds and speeds, cutting tables, 69, 70.
File, holding a, 270.
    selecting a, 269.
Files and filing, 269–280.
Filing out templates, 272–280.
Finishing lathe work, 127, 128.
Fitting brasses to their boxes, 284, 285.
    connecting rods, 314–324.
    cylinders, 288–297.
    link motions, 285–288.
Flanges, fitting to boilers or flanges, 313.
Flat drill, experiments with, 204.
    drill for enlarging and truing out holes, 168–171.
    drill grinding, 202.
    drill, increasing the keenness of, by means of an ellipse, 202.
    drills, 201, 202.
Flute of a tap, spiral, 241.
    of a tap, volute, 241.
Flutes for small taps, 243.
Foot lathe, 411–416.
    power lathe of W. F. & John Barnes Co., 414–416.
Force pumps, 488, 489.
Forging and hardening lathe tools, 25.
    tools, 220–222.

Forms, special, of lathe tools, 54, 55.
Four-flute tap, 246, 247.
Fox's pillar-shaping machine, 440.
France, pitches of threads of screws used in, 102.
Franklin Institute, standard for screw threads adopted by, 243.
Friction clutch lathe, 424, 425.
"Front Tool," 26, 28.
    Tool, for brass work, 50–53.

GAS taps, the taper of, 241, 242.
Gauge for grinding and setting screw tools, 92, 93.
    for testing the angles for threading tools, 90–92.
Gauze joints for high temperatures, 313.
Gear of Hendey-Norton lathe, 418–421.
Girder, wear of, 232.
Gisholt Machine Co., turret lathe of, 426–428.
Grain of properly forged tool steel, 226.
Graver, the, 154–159.
    the application of, 155–157.
Grinding, holding a tool in, 352, 353.
    pointed tools, 355.
    reamers, 172–174.
Grindstone and tool grinding, 347–359.
    for engravers' plates, 349.
    for tool grinding, 349.
    of uneven surface, using a, 355.
    the face of a, for flat surfaces, 352.
    truing device for, 350, 351.
Grindstones, different varieties of, and their qualities, 347, 348.
    fitting, with traversing rests, 347.
    should be run true, 349.
    the surfaces of, 348.
    use of, 347.
Grooving tool, 61–63.
    tool for brass work, 45.
    tool for wrought-iron or steel, 44–46.
    tools, 44–47.
Gun metal, 237.

HALF-ROUND bits, 166–168.
    Hand chasing, 104–107.

INDEX. 499

Hand nut for foot lathe, 413.
    rest of lathe, 412, 413.
    stocks, dies for use in, 251-255.
    taps, 248, 249.
    tools, 25.
    turning, 151-163.
    turning brass work, 159, 160.
Hardening, 227.
    American chrome steel, 226.
    and tempering tools, 222-227.
    cast-steel, 228.
    lathe tools, 25.
    springs, 227-229.
    taps, 244-246.
Hard metals, drilling, 205, 206.
    saw blades, to cut, 304.
Hardness in tools, sacrificing by overheating, 225.
Headstock of foot lathe, 411.
Heat, proper, in forging taps, 238, 239.
Heel tool, 157-159.
    tool, hardening, 158.
Height of the cutting edge of a tool, relation of, to the work, 40-42.
Hendey Machine Co., pillar-shaper of, 442, 443.
Hendey-Norton lathe, 416-421.
        lathe, gear for, 418-421.
Holders, boring tool, 82, 83.
    tool, 55-63.
Holes, small, machine for drilling and tapping, 459.
    to enlarge and true out, 168.
Horton's two-jawed chuck, 131.
Hubs of unequal height, to mark off the distance between the centres of two, 379-381.
Hurlburt-Rogers Machine Co., cutting off machine of, 435.
Huron Grindstones, 347.

INDEPENDENCE grindstone, 347.
Iron and steel, swelled by hardening, 228.
    case-hardening, 229, 230.
    planers, 445-452.
    planing machines, 445-452.
    side tools for, 47-50.
    wrought and cast, cutting speeds and feeds for, 69, 70.

JOINT, rust, 313.
    steam and water, 313,

Joint, the ground or scraped, the best, 313.
Joints, gauze, for high temperature, 313.
    of canvas or duck coated with red lead, 313.
    ordinary, 313.
    red lead, 313.
    rubber, 313.
Journal boxes, brass for, 237.
Journals, fast-running, Babbitt metal for, 237.

KEYS, reversed, 329-331.
    Keyway cutter, Morton Mfg. Co.'s, 466-468.
Keyway cutting machines, 465-468.
    marking out a, 402.
    or slotting drills, 206-211.

LATHE chucks, 131-135.
    cutting tools, classification of, 25.
    dogs, carriers or drivers, 116-119.
    foot, 411-416.
    foot power, of W. F. & John Barnes Co., 414-416.
    hand, expert, 25.
    Hendey-Norton, 416-421.
    in which the spindle for the dead centre is square, 426.
    screw-cutting engine of New Haven Manufacturing Co., 421-424.
    simplest form of the, 411-416.
    the chief of all machine tools, 411.
    the importance of, 25.
    tools, cutting surfaces of, 84.
    tools, forging and hardening, 25.
    tools, special forms of, 54, 55.
    turret, of the Gisholt Machine Co., 426-428.
    work, boring tools for, 71-83.
    work, centring, 122-127.
    work, finishing, 127, 128.
    work, speed and feed in, 64, 65.
Lathes and planing machines, cutting tools for, 25-63.
Leaky plugs, to rivet to their cocks, 304-308.
Left-hand thread, wheels necessary to cut, 102.
Lever arms, boring, 144, 145.
Line, a straight, to divide into a number of equidistant points, 367.

Line-shafting, setting in line, 331–337.
Line, straight, to divide into two equal parts, 367.
Lining or marking out work, 360–410.
 or marking out work, tools employed in, 365, 366.
 out a double eye, 383–389.
 out connecting rod, 396–408.
 out work, accuracy required in, 361.
 out work, importance of, 360.
 up brasses, 324.
Link motions, fitting, 285–288.
Links or levers, boring, 144, 145.
Live centre of foot lathe, 411, 412.
Liverpool grindstones, 347.
Long continuous cuts, tool for, 33.

MACHINE steel, hardening, 228.
 tool, what it is, 411.
 tools, 411–473.
 tools are cutting wedges, 29.
 using a boring bar should not stop while the finishing cut is being taken, 180, 181.
Machines, drilling, 458–465.
Mandrils or arbors, 120–122.
Marker out, qualifications necessary in a, 361–363.
 out, tools employed by the, 365, 366.
Marking holes at a right angle, 381–383.
 off a crosshead, 381.
 or lining out work, 360–410.
 out a cone pulley, 408–410.
 out a cubical block, 369–373.
 out a keyway, 402.
 out an eccentric, 389–396.
 out an engine guide-bar, 373–378.
 out a rod end, 397.
Mark off the distance between the centres of two hubs of unequal height, to, 379–381.
 out an ellipse, 363, 364.
Measuring work before marking out, 369.
Melling grindstones, 347.
Metal surfaces, wear of, 230, 231.
Metals, hard drilling, 205, 206.
 to be cut, influence of, on shape of tool, 26.

Milling bar and cutters, 339–346.
 cutter, 339–346.
 cutters, small size, 344.
 machine, importance of, 338.
 machine, plain, Brown & Sharpe, 453–455.
 machine superseding the shaping machine, 411.
 machine, universal, Brown & Sharpe, 455–457.
 machines, 453–458.
 machines and milling tools, 338–346.
 machines, the cutters of, 64.
 more rapidly executed than planing, 411.
 tools and milling machines, 338–346.
Mixture of metals, 237.
Monitor lathe, Warner & Swasey's, 424–426.
Morse Twist-Drill Co., thread upon the taps of, 344.
Morton Manufacturing Co.'s keyway cutter, 466–468.
Morton's shaping machine, 444, 445.
 shaping machine, novel and interesting features of, 444, 445.
Mushet's "special tool steel," 220.

NATIONAL bolt cutter, 471–473.
 New Castle grindstones, 347.
New Haven Manufacturing Co.'s screw-cutting engine, lathe of, 421–424.
Nova Scotia grindstones, 347.

OBLONG holes, drilling, 206.
 Ohio grindstones, 347.
Oil-hole for a strap for a connecting or side rod, 322.
Overheating tools in hardening, 225.

PANELLING and dovetailing machines, Boult's, 210.
Pening, 282–284.
Pillar shaper of the Hendey Machine Co., 442, 443.
 shaping machine, Fox's, 440.
Pin drill, employing as flat-bottomed countersink drill, 212.
 drills, 211, 212.
 tempering, 211, 212.
Piston, boring, to receive the piston rod, 145.

INDEX.

Piston pumps, 489–494.
 ring, expanding chuck for holding, 150.
 rings, 146–150.
 rings, cast-iron, wear of, 230.
 rings, inside diameter or bore of, 147.
 rings, turning, 147.
Pistons and rods, turning, 145, 146.
Pitch of screw, a coarse tool for cutting, 84, 85.
 of screws, to calculate the changed gear wheels for, 94–102.
Planers and shapers, 435–452.
Planer tool, a, 42, 43.
Planing machines, cutting speed of, 64.
 machines, cutting tools for, 25–63.
 machines, iron, 445–452.
Plug tap, 241.
Plugs, leaky, to rivet to their cocks, 304–308.
Points through which the curve of an ellipse may be drawn, to find, 364.
Pratt & Whitney tap, 248, 249.
Prentice Bros., drilling machine of, 462–465.
Principles affecting the shape of cutting tools, 26, 27.
Pulley, cone, to mark out a, 408–410.
Pulleys, wheels, etc., to calculate the speed of, 474–476.
Pumps, 486–494.
 force, 488, 489.
 piston, 489–494.
 suction, 486–488.

QUINT'S six spindle turret drilling machine, for drilling and tapping small holes, 459.

RADIAL drill or radial drilling machine, 465.
Rake, bottom and side, of a tool, what dependent on, 40.
 effect of having little in a tool, 30, 31.
 front and side, a tool with, 34, 35.
 front, in a tool, effect of, 34.
 in cutting tools, 27–29.

Rake of a tool, and diameter of work, relation of, 40.
 side, and front, in a tool, 33.
 side, in a tool, 31–33.
 side, in a tool, effect of, 34.
 top, effect of a great, in a tool, 29, 30.
 top, in a tool, 35.
 top, in a tool, effect of, 31.
Reamer, adjustable for small work, 175–177.
 considerations in determining the form of, 171, 172.
 method of grinding, 172–174.
 standard, great advantage of, 174.
 the, to maintain to standard diameter, 174.
Reamers, 171–177.
 adjustable, 174–177.
 and bits, 171.
 shell, 176, 177.
Re-centring work which has already been turned, 126, 127.
Reciprocating and revolving surfaces, wear in, 234, 235.
Red lead joints, 313.
Reversed keys, 329–331.
Riveting work by shrinking it, 308–312.
Rivet leaky plugs to their cocks, to, 304–308.
Rod end, marking out a, 397.
Rotary engines, difficulty of the success of, an account of the inequality of wear in side or disc surfaces, 233.
Round-nosed tools, 36, 37.
 tools, cutting edge on, 36.
Roughing-out brass work, tool for, 159, 160.
 hand-turned work, 154.
 tool for, 33.
Rubber joints, 313.
Russell Tool Co.'s drill chuck, 134.
Rust joint, 313.

SAW blades, hard to cut, 304.
 Scale in hardened steel, 226.
Scraped surfaces, 297–301.
Scraper, best form of, 301.
Scrapers and scraping, 280, 281.
 for brass work, 160–163.
Screw-cutting engine lathe of New Haven Manufacturing Co., 421–424.
 gear, compound or double, 97.

Screw-cutting machine a form of lathe, 411.
  machine, a small, illustrated and described, 429–432.
  machine, great advantages of, 428, 429.
  machine of the Brown & Sharpe Mfg. Co., 432, 433.
  machines, 428–434.
Screw thread, the United States standard, 88–90.
  thread, the Whitworth, 88, 89.
  thread, a double, to cut, 102–104.
  threads, in use in the United States, the shapes of, 87–89.
  threads, standards for, 243, 244.
  threads used in France, pitches of, 102.
  to cut, by hand in the lathe, 104–107.
  cutting tools, 84–115.
  tools, gauge for grinding and setting, 92.
Screws, to calculate the change gear wheels to cut the pitch of thread for, 94–102.
Scribing block, 268.
Sellers, Wm., & Co.'s experiments with a flat drill, 204.
Sellers, Wm. & Co., slotting machine of, 468–470.
Setting line shafting in line, 331–337.
Shafting, line, setting in line, 331–337.
Shape of cutting tools, 26.
  of work, importance of, in hardening, 227.
Shapers and planers, 435–452.
  traverse, 440–442.
Shaping machine being superseded by the milling machine, 411.
  machine, Morton, 444, 445.
  machine, Morton, novel and interesting features of, 444, 445.
  machines, tool holders for, 60–62.
Sheet-iron, cutting out holes of a large diameter in, 217.
Shell reamers, 176, 177.
Shrinking, riveting work by, 308–312.
Side planer, open, Detrick & Harvey Machine Co., 451, 452.
  rest tools, 25.
  rest tools, classifications of, 26.
  tool, 26.
  tool for brass work, 53, 54.
  tool for small work, 48, 49.
  tools for iron, 47–50.
  tools, left-handed and right-handed, 48.
Sizing die for finishing taps, 242, 243.
Skin of iron or brass castings, and iron or steel forgings, hardness of, 342.
Slate's sensitive drilling machine, 459–462.
Slide-rest for foot lathe, description of, 413, 414.
  rest of cutting off machine, 434.
  valve, considerations in setting, 477, 478.
  valve, to set a, 477–485.
  valves, cast iron, wear of, 230.
Slot drill, application of the principles of the action of, 210.
Slotting drill, labor saved by, 210.
  drills, tempering, 209.
  machine, 468–470.
  machine of Wm. Sellers & Co., 468–470.
  machine tools, 191–195.
  or keyway drills, 206–211.
Small holes, drilling machine for drilling and tapping, 459.
Soldering liquid, 237.
Solders, 237.
Speed and feed, cutting, 64–70.
  of wheels, pulleys, etc., to calculate, 474–476.
Speeds and feeds, cutting tables of, 69, 70.
Spindle of a cutting off machine, 432, 433.
Spring tool, 26.
  tool for finishing sweeps, curves, and round or hollow corners, 46.
  tool, the top face of a, 47.
Springs, hardening, 227–229.
  the steel for, 227, 228.
Spur centre, the wood turner's, 119, 120.
Square-nosed tool, setting and feeding, 38.
  tools, 37–43.
  tools, what used upon, 37, 38.

INDEX.

Square, the, 267.
Steadying device and tool holder, 59.
Steam and water joints, 313.
   valves and cylinder posts, marking out, 406-408.
Steel, cutting speeds and feeds for, 69.
   for taps, 238.
   grooving tool for, 44-46.
   hardening and tempering, 222.
   tool, 25, 219, 220.
Stock and cutter, in cutting out holes of a large diameter in sheet-iron, 216, 217.
Straight line, to divide, into two equal parts, 267.
   line, to divide into a number of equidistant points, 367.
Strain on a tool, 29.
   upon a tool in cutting, 30.
Straw color in tempering tools, 225.
Suction pumps, 486-488.
Surface plate, to make a, 301-304.
Surfaces scraped, 297-301.
Sweetland chuck, the, 133, 134.

TABLES of cutting speeds and feeds, 69, 70.
Tailstock of foot lathe, 411-413.
   of lathe capable of being moved or fed crosswise of the lathe, 426.
Tap, a three flute, 246, 247.
   finishing the thread of, by passing through a sizing die, 242, 243.
   hand and machine-made, turning the taper of 241, 242.
   taper, 241.
   the nut, 240.
   the plain part of a, 242.
Taper in taps, 241, 242.
   plug, 241.
   tap, 241.
   tap, the proper taper for, 239.
Taps and dies, 238-255.
   for holes requiring to be exact in diameter, 241.
   for holes to be tapped deeply, 242.
   for ordinary work, 241.
   for use in machines, threads of, 240, 241.
   forging, 238.

Taps, gas, the taper of, 241, 242.
   heating, for hardening, 244-246.
   heating in forging, 238.
   small, flute for, 243.
   standard, screw thread for, 243, 244.
   steel for, 238.
   taper in the diameter of the bottom of the thread, 241.
   threads of, 239, 240.
   threads of, finishing, 239.
   three-fluted and four-fluted, 246-248.
   with thread on small end of taper, 242.
Temper for drills, 204.
Tempering and hardening tools, 222-227.
   a tool at or near the cutting edge only, 223.
   cutters, 217, 218.
   pin drills, 211, 212.
   slotting drills, 209.
Templates, filing out, 272-280.
Thread, internal screw, tool for cutting, 87.
   left-hand, wheels necessary to cut, 102.
Threading tools, gauge for testing the angles of, 90-92.
Threads of screws, cutting small and thick, 84, 85.
   of screws, coarse, cutting, 84, 85.
   of screws in use in the United States, shapes of, 87-89.
   of taps, 239, 240.
   of taps, finishing, 239.
   of taps of small size, finishing, 239.
   on wrought-iron or steel, tool for cutting, 86.
   outside V, in brass work, tool for cutting, 86.
   V, in iron, tool for cutting, 86.
Tit drill, 202.
Toggle of Warner & Swasey's lathe, 425, 426.
Tool feed, rake of, 40.
   for roughing-out and long continuous cuts, 33.
   grinding and grindstone, 347-359.
   hardening and tempering, 222-227.

INDEX

Tool holder and steadying device, 59.
  holder for planing machine tools, 61–63.
  holder, Woodbridge's, 57–59.
  holders, 55–63.
  holders for a shaping machine, 60–62.
  post of Hendey-Norton lathe, 421.
  steel, 25, 219, 220.
  steel, Mushet's, 220.
  with a combination of front and side rake, 34.
Tools and tool holders, Woodbridge's, 57–59.
  angles of, 39.
  cutting, for lathes and planing machines, 25–63.
  employed by the marker out, 365, 366.
  for cutting various screw threads, 86, 87.
  forging, 220–222.
  for use in slotting machines, classification of, 181.
  machine, 411–473.
  round-nosed, 36, 37.
  screw-cutting, 84–115.
  square-nosed, 37–43.
Top rake, application of, to a boring tool, relation to the strain, 74, 75.
Traverse shapers, 440–442.
Turning cranks, 140.
  eccentrics, 136.
  hand, 151–163.
  pistons or rods, 145, 146.
Turret lathe of the Gisholt Machine Co., 426–428.
  lathes, 426–428.
Twist drills, 196–201.
  drills, cutting edges of, 197–200.

UNITED States standard for screw threads, 243.
  States standard screw thread, 88–90.
  States, the screw threads in use in, 87–89.

VALVES, brass for, 237.
  Vise clamps, 280, 281.
Vise work, 256–313.

WARNER & Swasey's universal monitor lathe, 424–426.
Water and steam joints, 313.
Wear arising from motion in one continuous direction, 230.
  greater in revolving than in reciprocating surfaces, 234, 235.
  inequality of, in revolving side or disc surfaces, 233.
  of metal surfaces, 230–236.
Wheels, pulleys, etc., to calculate the speed of, 474–476.
Whitworth or English standard for screw thread, 244.
  or English standard taps, flutes for, 246, 247.
  screw thread, 88, 89.
  Sir Joseph, lathes of, 41.
  stocks and dies, 253.
  taps, flutes and teeth, 248, 249.
Wickersly grindstones, 347.
Woodbridge's patent tool and tool holder, 57–59.
Wrought-iron, case-hardening, 229, 230.
  cutting speeds and feeds, 69.
  grooving tool for, 44–46.
  screw-cutting tools for, 84.

YELLOW brass, 237.
  brass for castings, 237.

Printed in the United States
61368LVS00005B/48

McAlester Public Library
401 North 2nd
McAlester, OK 74501
a branch of the
Southeastern Public Library
System of Oklahoma